❧ CHARLES SUMNER SLICHTER

The Golden Vector

CHARLES SUMNER SLICHTER

The Golden Vector

MARK H. INGRAHAM

The University of Wisconsin Press
Madison, Milwaukee, and London

Published 1972
The University of Wisconsin Press
Box 1379, Madison, Wisconsin 53701

The University of Wisconsin Press, Ltd.
70 Great Russell Street, London

First printing

Printed in the United States of America
Kingsport Press, Inc., Kingsport, Tennessee

ISBN 0-299-06060-8; LC 79-176412

To Katherine

Who for nearly half a century
has shared with me the friendship
of the Slichters.

ᔍ Contents

List of Illustrations   ix

Acknowledgments   xi

1   Introduction   3

2   Ancestry and Youth   7

Birth and Ancestry 7. Boyhood 11. Northwestern University 16.

3   Department Member   21

Teacher 22. Textbooks 38. Professional Societies 39. Chairmanship 39.

4   Scientist and Engineer   55

Scientist 56. Engineer 69.

5   University Citizen   102

Committees 102. The University Club 129. The Occasional Speaker 132.

6   The Graduate Deanship   139

The Research Committee before 1920 139. Appointed Dean 143. Administrative Colleagues 145. Relations with Students 150. Younger Faculty Members 157. Rules and Their Enforcement 165. Relations with Colleges of the State 167. Carl Schurz Professorship 169. Association of American Universities 170. A Presidency Declined 172. Pride in the University of Wisconsin 172. Defense of Research Budget 173. Refusal of Foundation Gifts 175. Wisconsin Alumni Research Foundation 176.

7   Business and Civic Affairs   194
Business 194.   *Civic Activities 204.*

8   Social Life   213
The Madison Literary Club 213.   *Town and Gown 214.*

9   Essayist   224

10   The Bureau Drawer   232
Nature and Title of This Chapter 232.   *Health 233.*
Do-It-Yourselfer 233.   *Robberies and Thefts 235.*   *Two*
Versions of One Story 236.   *Charities 239.*   *Respectful*
Iconoclast 240.   *A Letter from Turner 241.*   *Gathered from*
Others 243.   *Tributes 243.*

11   Family Life   248
Mary Louise Byrne 248.   *1890–1909 253.*   *The Year Abroad*
259.   *Long Lasting Family Interests 264.*   *1910–1934 284.*
Retirement 289.

Chronology of Charles Sumner Slichter   299

Index   301

❧ Illustrations

Frontispiece

Charles Sumner Slichter

Page 73

List of groceries ordered for a survey camp

Following page 130

Charles S. Slichter, about age seventeen
Mary Louise Byrne before her marriage
Slichter as a young faculty member
Dean Slichter
Mrs. Slichter, when her husband was Dean of the Graduate School
The Slichter boys: Louis, Allen, Donald, and Sumner, about 1902
"The Slichter Boys" being given citations, May 3, 1957
The Slichter home on North Frances Street
Sumner Slichter in his baby carriage

Page 284

Slichter's checks

ᔫ Acknowledgments

There are so many persons to whom I am indebted in connection with this biography that my thanks, though sincere, may seem dilute. I am cheered by the certainty that they would find each other congenial. Some of this goodly company I mention below.

Edwin B. Fred, President Emeritus of the University of Wisconsin, not only urged me to write a life of Charles S. Slichter, but furnished me with many documents which he had ordered and annotated. He frequently talked with me about the dramatis personae of this story, particularly about its lead role. He and Ray M. Stroud arranged for my access to the Minutes of Town and Gown and gave me permission to quote from them. President Fred read the chapter on the graduate deanship and gave me his point of view as Slichter's successor in that position. All of this merely added to the long list of his kindnesses for which I am grateful.

I received encouragement and wise counsel from Fred H. Harrington when he was president of the University of Wisconsin and from Robert L. Clodius while he was vice-president.

Slichter's sons, as well as their wives and children, helped me in every possible way. They allowed me to quote from their correspondence. They furnished me with letters, photographs, and memories to supplement the Slichter files in the Archives of the University of Wisconsin—files made rich by much that they had previously deposited therein. The sons read and commented on the next-to-the-final draft of the manuscript.

The staff of the Archives—including Jesse E. Boell, the director when this study started; and his successor, J. Frank Cook; and their secretary, Mrs. H. Ruth Thayer—gave me effective aid as well as made pleasant my work in their realm.

I wish to thank the persons who made my visit to Garden City, Kansas, pleasant and productive: Mr. Foster Eskelund, of Deerfield, the leading spirit of the Kearny County Historical Society, guided me around the sites of Slichter's activities, meanwhile pouring forth a rich flood of interesting information. Mr. and Mrs. J. O. Carter and Mrs. Dale Vaughn discussed the history of Garden City with me and opened the Finney County Historical Museum so that I could examine copies of old local newspapers. Mr. Ray Calihan, the attorney of the Finney County Water Users Association, and the secretary, Mrs. Luava Golightly, not only made the Association's records available to me but furnished work space and Xeroxing facilities. The three days in Garden City and vicinity were among the most illuminating which I have had in years.

The work of three former graduate students of the University of Wisconsin was very useful to me: David F. Allmendinger, Jr., summarized the Slichter material in the Archives and made a bibliography of Slichter's writings. He also wrote vivid accounts of interviews with a number of persons who had known Slichter well. Charles A. Culotta spent a summer studying the growth of financial support of scientific research at the University of Wisconsin. His unpublished report on the subject contained much of value for this biography. Edward H. Beardsley's *Harry L. Russell and Agricultural Science in Wisconsin* gives an account, centered on Russell, of many of the events described in this book.

W. J. Condo, business officer at the University of Manitoba, and the Metropolitan Toronto Central Library furnished me valuable material in connection with Slichter's work on the water system of the City of Winnipeg.

The Reverend Paul V. Hoornstra, Rector of Grace Episcopal Church of Madison, and his staff not only allowed me to consult the records of the church but helped me to find the material I needed.

Anyone writing about the past of the University of Wisconsin uses and admires the history (1848 to 1925) of the University by Merle Curti and Vernon Carstensen.

I am indebted to Robert Taylor for permission to quote from his unpublished manuscript, "The Birth of the Wisconsin Alumni Research Foundation"; to George A. Evans for permission to quote from letters of his father, Judge Evan A. Evans; to Jackson Turner Main for permission to include a letter from his grandfather, Frederick Jackson Turner; and to Mrs. Ray Owen for permission to quote from letters of her husband.

Charles Scribner's Sons kindly allowed me to quote from Warren

Weaver's autobiography, *Scene of Change;* and Weaver permitted me to use his letters and addresses.

The following list indicates where certain of Slichter's essays were first published. When it was deemed necessary to secure permission to quote from them, it has been readily granted.

"Industrialism," *Popular Science Monthly,* October 1912;

"The Principia and the Modern Age," *American Mathematical Monthly,* August-September 1937;

"The Self-Training of a Teacher," *Journal of Engineering Education,* October 1931;

"Galileo," *American Scientist,* April 1943;

"Polymaths: Technicians, Specialists, and Geniuses," *Sigma Xi Quarterly,* September 1933.

All of these, except "Galileo," were later republished in *Science in a Tavern.*

Sumner Slichter's "Technology and the Great American Experiment" was published as a pamphlet by the University of Wisconsin Press, 1957.

I wish to thank Mrs. Alice Schroeder, who, with great good cheer, typed the various drafts and the final manuscript of this book; Miss Catherine Goddertz, who helped search the Archives—perceptively selecting material to be studied—who verified many quotations and references, and helped edit the final draft; and Mrs. Helen Guzman, who, as frequently before, has given me invaluable aid, organizing the project and reading with careful scrutiny most of the text—improving many passages and insuring that, if I made an odd use of a word or employed an unorthodox grammatical structure, it was done consciously. I believe all three enjoyed becoming acquainted with Slichter.

We have taken the liberty of correcting many typographical errors in letters and other documents, especially in carbon copies where it is likely that the unavailable original had been corrected.

Undoubtedly this volume contains errors, for some of which I am responsible; but this responsibility is also shared with others—perhaps chiefly with the hero of this tale. But, although there are many who have aided and abetted me, my chief help also came from Charles Sumner Slichter, who filled me with admiration and kept me amused, interested, and bewildered.

Mark H. Ingraham

Madison, Wisconsin
May 1971

CHARLES SUMNER SLICHTER

The Golden Vector

Chapter 1 *Introduction*

C HARLES SUMNER SLICHTER had a fascinating and colorful personality. It was a joy to know him. This is the chief reason why his biography should be written, but there are others. He was an important figure in the life of the University of Wisconsin. He was warm in affection and artistic in anger. He knew interesting people and was interesting to them. He was *sui generis* yet representative of some things we are in danger of losing, including "sui generisity." His memory should be perpetuated by all the means at our command.

Heretofore much of my writing has been an effort to affect the future. I have now turned to the easier task of affecting the past. The past may be misrepresented, or carelessly represented; but, in addition to these reprehensible treatments, we choose what we wish to remember or what we try to forget. The past we live in is more malleable than the future.

In recollecting the past we think of important events and we cherish the memory of interesting people. There are a great many people whom I respect and find congenial. The number who are truly interesting is more limited. We can increase the proportion of interesting people whom we know by choosing to be acquainted not only with some contemporaries but also with a selection from the past. Slichter did this. We can do so by knowing him.

There are many good reasons why someone other than I should have written his life. If a historian had written it, it would be more professional. If it were a Ph.D. dissertation, it would be more scholarly. Few elder men can make a reputation, many can lose one—but I have neither an ambition to be known as a biographer nor a reputation in that field to be destroyed. I have tried to be accurate, but I have had no intention of mastering either the craft or the paraphernalia of producing footnotes. Perhaps the chief justification for me, rather than someone else, to have undertaken the work was that I was urged to do so by Edwin B. Fred, the beloved President Emeritus of the University of Wisconsin, who—although he knows one is busy—still wishes to be doubly sure that none of his friends falls into the clutches of the employer of idle hands.

I agreed to the task largely because I had known Slichter well and had found his company both stimulating and good fun; and I was certain that prolonging my acquaintance with him, even to the years before I was born, would be enjoyable. If I had realized how much work it would be, I might not have had the temerity to consent to writing this biography; but if I had known how much fun it would be, I would have asked for the privilege. It would have spared me the work, but deprived me of the fun, if Slichter had taken the 1938 suggestions of "Billy" Kies and Max Mason that he write his autobiography.

My first purpose is to portray Slichter's personality and my second (but not secondary) is to describe his accomplishments. These accomplishments were large; moreover, to an unusual extent they were the corollaries of his personality—seldom in any gathering of which he was a part was there an equally striking individual.

The goals of being a chronicler and of being a portrayer are not separable, for it would be a mistake to think of Slichter as a presence divorced from action; rather, his was a style of life expressed through action. His interests were manifold. As a scientist, he wrote one fundamental paper and a number of creditable ones. He became—in connection with the flow of underground waters—an operating engineer, a sought-after consultant, and an expositor. He was one of the chief architects of the Wisconsin Alumni Research Foundation, which has to date granted over fifty-five million dollars to the support of research at the University of Wisconsin. He was the dominant member in a host of committees whose work was a part of the warp and woof of a great university. In particular, he formed and for seventeen years led its Research Committee. He was an atypical but successful administrator. (Perhaps success as an administrator is itself atypical.) A humanist, he wrote magnificently. Yet I believe he would have preferred to be remembered first and longest as a teacher.

He carried on many diverse activities and had a remarkable ability to turn his attention from one to another. An attempt to give a chronological account of his life would result in such a kaleidoscope of topics as to thoroughly confuse the reader. His fiancée wrote to him: "You are probably wearing everyone out," a guess which she had ample opportunity to verify. Hence, in most cases each chapter deals with some phase of his life over a protracted period. These phases persisted simultaneously. A short chronological guide is given as an appendix so that one can place an event in its proper time. Moreover, the next and the last chapters, taken together, provide a sequential picture of his family life.

I have avoided footnotes, both those that prove one a meticulous scholar and those that are more interesting than the text, but which in a serious work are likely to be put in a secondary position. However, a file, with page references, containing copies of most of the material used and citations, is in the University of Wisconsin Archives.

Memory, although unreliable as to detail, makes it possible to vivify the facts of record. The use of memory suggests one question I had to settle: Should I write in the first person singular? Since this biography is conditioned by the fact that I was a student and a colleague of Slichter's and participated, sometimes actively, in many of the events of the last quarter-century of his life, I decided that it was preferable not to assume the mask of third-person objectivity.

Another problem, and I am sure I have not been totally successful in dealing with it, was how to make clear Slichter's abundant and ever-ready humor without making him seem to be a "funny man." He was a man of definite purposes and deep feelings, some of which he hardly dared show without a touch of banter. In life, his wisdom gained through its dress of wit; in the retelling, it may lose. And it was not only wit he liked, but also emphasis. He could explode naturally, but also deliberately, to add strength to his utterances. He would not sacrifice effect to trivial accuracy, let alone mere consistency. His positiveness in making statements now difficult to verify is a comfort to one who may not always have qualified sufficiently his conjectures about Slichter himself.

My earliest choice of title was: "HELL-ROARING CHARLEY—The Story of a Cultivated Mind." I still like this title but it would not do, for it overemphasized two of the many facets of Slichter's personality. Moreover, his sons could not understand how he earned the nickname; and I fear there were some whose rapid retreat never allowed them to sample the intellectual fruit of his tillage.

In a statement contrasting the worlds of science and of personality, Slichter wrote: (In his essay on "Polymaths," included in *Science in a Tavern*, he puts these ideas into the brain of Heaviside.)

> . . . *there is a greater world, the*
> **World of Personality.**

*The room of this world is not the frigid space of
science, but the warm continuum of the Great
Presence.*

. . .

*Also, for each personality, there is a Time's Arrow
—a Golden Arrow—trued and checked against
the frame of Truth, Beauty, Virtue, and the trend
of the arrows we call Purpose.*

. . .

*The Great Presence is mindful of the arrow, for as
that golden vector moves and orbits in a spiral
of life, there exists a Best Way and many less perfect ways, and the Great Presence would guide
and shield that orbit in the way that is best and
lead it more and more to its own perfect purpose.
This is the spiritual creed of evolution, and it holds
alike for galaxies and for men.*

From this, the subtitle "The Golden Vector" is taken.

ᴥ *Ancestry and Youth*

Birth and Ancestry

W HEN we begin life our first screams as infants are merely an instinctive groping for blasting words." This autobiographical remark describes an event which occurred probably on April 16, 1864, although, as an exception to other records, the certificate of baptism (there was no birth certificate) gives the birth date of Charles Sumner Slichter as May 16, 1864.

The infant's forebears were more devout than profane.

Sometime, not earlier than 1935, he wrote two accounts of his ancestry—one entitled, "Two Hundred Years of Pioneering, as told by Charles S. Slichter," and a shorter version, "Two Hundred Years of Pioneering, as told by Pat O'Fessor." (He was fondly called " 'fessor" by his grandchildren.)

It is not clear whether the shorter of these two manuscripts is an abridgment of the other or whether he had rewritten it as new facts were dug up. In either case we cannot do better than to quote the story as he told it. Since practically everything in the shorter version is contained in the longer, I have used the longer one, with some cross-checking, and have done my own cutting:

~ *The name Schlichter means mediator, arbitrator, or planisher, contrasting with* Richter, *which means judge in the legal or official sense. The form* Schlichter *is high German, the form* Slichter *is Swiss dialect. The job of a* Slichter, *if true to his name, is to make smooth for others the rough places along their way. The job of a* Richter *is to straighten, that of a* Slichter *is to make smooth. The name is not very common any where but is most common in Switzerland and Alsace, and is not uncommon in the Palatinate, especially near Heidelberg and Frankenthal. In the United States the name seems to have been most common*

in Franklin County, Pennsylvania, (County seat, Chambersburg) and
in Montgomery County, Pennsylvania, (County seat, Norrisberg).

The first migrants of that name came to America in the eighteenth
century and settled, for the most part, in the two counties named. All
were apparently strong Protestants of various sects, chiefly Calvinists
of which there were many branches or divisions. None seemed to have
been Lutherans in the usual sense of the term. Many were "Swiss
Brethern" or Swiss Mennonites, a sect opposed to bearing arms or
taking oaths or accepting public office. They believed that people
could get along without law (except the ten commandments) if they
were sincere in their Christian belief and they thought that people
should settle all differences by Schlichters and not by Richters. Extreme branches of the Mennonites—such as the Amish sect—wore
special dress like Quakers and were exceedingly clannish and secluded
and unsocial in relation to the outer world. There seems to have been
no Schlichters in this extreme group.

He then lists Slichters, or those with variations of that name, who
came from Rotterdam to Philadelphia between 1727 and 1808. His
account of the direct line continues:

~ . . . Christian Schlichter, my great grandfather, is not on the list.
He must have entered at Baltimore or elsewhere or perhaps the ship
list containing his name has been lost.

Many Schlichters settled near New Hanover and Upper Hanover,
Montgomery County, and many are buried in the Perkasie Meeting
House Cemetery near Hilltown. It is obvious that several unrelated or
very remotely related families of that name settled in Pennsylvania,
and it is clear that they were of many different Protestant sects. Some
of the early Christian names of the Schlichters are interesting. Among
these are Gallus, Jost, Nicholas, Andrew, Christopher in Montgomery
County, but the favorite names of the Swiss Schlichters in Franklin
County, Pennsylvania and Frederick County, Maryland, were John,
Henry, Samuel, Jacob, Benjamin and Christian. There are almost no
James, Matthews, Michaels, Isaacs, Solomons, or Josephs.

The spelling of the name took on all possible forms, especially in
boat lists . . . , real estate titles, and probate records. English speaking
officials evidently had a hard time of it, for we find the forms Slighter,
Sleighter, Sleeghter, Schlighter, Schlichtder, Schlicker. The United
States census taker in 1790 spelled the name in all these ways in making his record in Montgomery County. In the Swiss community near
Chambersburg (Franklin County) the Swiss spelling Slichter was not
uncommon, this is also the form adopted by some but not many of

the Canadian settlers when they moved in 1806 from Pennsylvania to Waterloo County, Ontario.

There follows an account of the Mennonites of Switzerland seeking religious liberty, first in the Palatinate, then in Holland, then in Pennsylvania—where they had gone with the help of the Dutch—only later in many cases to move to Ontario. I quote again:

~ The Swiss migrants after reaching Pennsylvania soon prospered mightily. The Scotch-Irish settlers selected valley lands but the wise Swiss selected limestone lands on the uplands. Probate records show that after only twenty five years the Swiss had reached substantial wealth. Eventually they made Lancaster County the richest agricultural county in the United States and the Swiss Mennonites headed the list of the prize winners.

The journey from Pennsylvania in 1806 to Waterloo County, Ontario was not an easy one. There were no lack of roads, such as they were, but they were all "terrible" by any standard. The only fact known to me concerning the route taken to Ontario by my ancestors, is that they crossed the Susquehanna River just above its junction with the Juanita and then continued northwest to the Niagara River near the end of Lake Erie, then continued easterly past the Falls to about the present site of the village of Lewiston, New York where they were either rafted or ferried across the Niagara River, thence to an old Mennonite settlement "The Twenty," twenty miles from the river. From there it was about 20 miles to the Grand River and the 60,000 acres of the Beasley Tract. A map of the route taken will be found in "The Trail of the Conestoga" by Mabel Dunham, Toronto, 1942.

The Conestoga Wagon had already been invented by some genius or geniuses in Lancaster County. This vehicle, drawn by four or five horses was the conquering ship of valley and hill and forest—it was a transport not discouraged or blocked by steep grades, muddy trails or bridgeless rivers. It was the wonder transport of that day. In less efficient and less sturdy form, it later became the "covered wagon" of the plains. Indian trails along the Susquehanna and elsewhere became the highways of the pioneers and later became the routes of railways. Many scores of families from Franklin and Lancaster Counties made this trip. It is unfortunate that no contemporary account of the journey seems to have been preserved.

My grandmother, Elizabeth Bechtel, wife of John Schlichter, son of Christian Schlichter, was a Mennonite, although many of her clan were United Brethern or Moravian Church. Many of the Pennsylvania Bechtels were active in that sect. The United Brethern are noted for their printing presses, their music and their high educational and ethical standards. The eleven children of this John Schlichter took the liberty of attaching themselves to a great variety of sects. As eight of the children were girls it was natural that they should follow their husband's choice. But John Schlichter, Jr., graduated at Oberlin, Ohio, and spent his life as pastor of the Congregational Church at Sterling, Kansas, and Jacob Bechtel Slichter (my father) joined the church of England and, of course, became a member of the Protestant Episcopal Church in the United States. . . .

Why did the Pennsylvania Mennonite migrate to Canada? One reason, of course, was that the young people were always seeking good land. A major reason was their confidence in the British government which had scrupulously fulfilled the promise made to them over one hundred years previously in granting them religious freedom and exemption from military service and from taking oaths. During the Revolutionary War they as a class remained loyal to the British Crown although a few of them entered the armed services and one of them even became a member of Washington's staff. When the war of 1812 broke out, the United States Government treated them rough. The Mennonites could not be induced to take up arms, but they were forced into service as teamsters furnishing their own horses. After the peace the migration to Canada became a flood, for the British government had treated the Canadian Mennonites with exemplary justice. For service in Canada, the Mennonites were well paid $5 a day for the teams and $8 for four horse teams. . . .

In Waterloo County the Pennsylvania Dutch and the Scotch were perhaps the most numerous groups. The Calvinism of the two groups did not clash and the Swiss descendants gave a good example to the more serious Scots by their zithers, their singing and their love of flowers and festivals. Then there was another important group, Tories who had fled the States at the outbreak of the Revolutionary War. These furnished a marvelous leaven of culture to the other less sophisticated groups.

Waterloo County, Ontario, is a rich agricultural county and the settlers soon became prosperous and even wealthy. They built fine houses and even finer barns. They raised wheat, rye, barley, corn, oats, field turnips, field peas, clover, timothy, geese, turkeys, ducks, and chickens, and had cows, swine and many sheep. The homes were large and well-furnished. The Swiss settlers had no trouble in keeping up with their Scotch neighbors in thrift and bargaining. They were smart and they were keen traders.

From material which also includes facts about collateral lines we excerpt information concerning Slichter's direct ancestors.

~ I. *Daniel Schlichter was a native of Switzerland, probably of Canton Berne. His wife was a French Huguenot. He was a member of the "Swiss Brethern" or Mennonites.*

II. His son Christian Schlichter was born in Switzerland in 1761. . . . The exact date of his arrival in America is not known, but it was probably during the years 1780–1785, for by 1787 he had married in Franklin County and had one child. . . . Christian married Mary Wanderbach and had a family of five children. Some time during the early part of 1806 he, together with his family, moved to Canada and settled in Waterloo County, Ontario, a little north of Preston on the place known as "Greens Farm," where he lived until his death many years later. He was a Mennonite.

III. John Schlichter, eldest son of Christian, was born in Franklin County, Pennsylvania, on October 10, 1787 and died August 12, 1849. He married Elizabeth, daughter of Jacob and Elizabeth Bechtel. She was born November 24, 1795, and died February 1, 1876. They lived on a farm one mile east of New Dundee, Ontario. He was a Mennonite.

Boyhood

John Slichter had eleven children, the third of whom, Jacob Bechtel Slichter, married Catherine Huber. The couple moved to Galena, Illinois, about 1840, to Saint Paul, Minnesota, about 1850, to Chicago, Illinois, in 1869. All nine of their children were born in the United States, Charles Sumner being the youngest. Since he was about five when the family moved to Chicago, Saint Paul could have made little impression upon him.

His earliest memory of which we have a record concerns that period. In the spring of 1936 his granddaughter Sue had the measles and he wrote to her father, his son Louis, on April 18:

~ *We shall be anxious to hear about the progress of the measles. I still remember my own case when I was five years old. It was a very light case, but exceedingly "itchy." Tell Sue that I kept calling to everybody to "Itch me! Itch me!" thinking that it was their duty to help scratch, which the doctor had forbidden me to do myself. I assumed that his ruling did not apply to others.*

For a person who became so aware of his surroundings and so much a part of them, it is remarkable how seldom Charles Slichter dwelt on his boyhood in Chicago. For instance, we do not find references by him to any memory of the Chicago fire of 1871, although the family tradition is that the fire severely affected Jacob Slichter's finances. It is said that Charles was not a robust boy which may explain why, when he was nine, he spent the summer with relatives in Ontario. He later wrote of that summer as the happiest period of his boyhood. His vivid and delightful memories of that visit were recorded for his grandchildren in "Two Hundred Years of Pioneering."

~ *I visited my relatives in Canada for an entire summer in 1873. I saw no sign of poverty or want anywhere, either among the Swiss or the Scotch. The younger people for the most part migrated farther west to farms or to Michigan near Brown City, where there is a large transplanted colony from Waterloo County dating from about 1850. I have a high respect for those simple people. They got down on their knees every morning in family groups for prayer and for a few verses of scripture. They worked hard and laid up money. They lived well, for the earth yielded its abundance, and their gardens and orchards seemed to know that they were in the hands of the favored of the gods.*

I have a most vivid recollection of my four month visit among my numerous aunts and uncles in 1873. It was delightful and gave an enlarged outlook to a green and frail city boy. These people were all in the midst of an old-fashioned economy. Most of their needs were supplied from their own farms. To be certain that there would not be too many idle hours, there were home industries for both men and women. One farmer would have a small clay pit where in slack seasons he would make brick for his neighborhood. Another would have a lime kiln, another a small saw mill, and another, quite important, a large cider press to serve them all. There were still many spinning wheels to make yarn for their woolen hose and mittens and jackets, and there were a few looms on which light woolen sheets were woven—for no true Switzer could think of sleeping in winter between the linen or cotton sheets of the Yankees. The houses for most part were heated with wood stoves, but a few homes had hot air furnaces and were the wonder of the neighbors. The houses were not only equipped with the ordinary kitchen for preparing meals, but also had a grand or super kitchen for the manufacturing activities of the household. These were not lean-to's or summer kitchens of the Yankee type, but well built integral parts of the home. Here the butter was churned, the wool carded and dyed, the laundry work done, and loads of fruits and vege-

tables and meats and sausages prepared for home use. In a large fire-
place a huge copper kettle hung on a crane for use in making pickles,
etc., and especially for making tubs and tubs of apple butter in the
autumn. This last was a most delightful and strenuous ceremony. Barrels of cider were boiled down in the copper kettle, constantly stirred
night and day by means of a long wooden handle extending from the
rotating stirrer far into the room to a chair where the stirrers took turns
at the handle. The steam from the cooking cider, quartered apples,
and spices, filled the house with a never-to-be-forgotten odor. The
whole household and young people from the neighbors took part in
this work—coring apples, operating several pealing machines, etc.
while one played the zither and all sang—it was not work, it was a
party.

Almost no white sugar was used on the table or in the kitchen.
Maple syrup, maple sugar and honey were the substitutes in cakes and
cookies and on fruit. Rich cream was used in abundance, even for
salad dressing. Near the fireplace of the super kitchen was the brick
oven for the baking of bread and pies. The flour was not bought—it
was exchanged for wheat at the nearby mill. Soft soap came from the
row of lye barrels at the edge of the orchard—an impressive symbol
of the orderliness, and thrice-scoured cleanliness of the household and
all its contents, and of the linen and woolens and of those that wore
them.

Farm machinery was just coming in. Mowers, horse rakes and reap-
ers were used, but self binders were not yet. All grain and hay was immediately stored in the ample barn—they would not think of stacking
grain and hay outside. At threshing time the threshing machine stood
in the center of the barn floor and rotating horsepower driven by several teams stood outside at a specially prepared place. The field peas
were threshed on the barn floor with flails, and the swinging of the
flails in time with a Swiss song was not work—it was a festival.

One of my grandparents was still living in 1873,—Elizabeth Bechtel
Slichter. She lived in a wing of the old homestead, the main portion
of which was occupied by her daughter Elizabeth and family. She was
the patriarchess of the community and, of course, looked up to by
all. She was not strong, but she was alert and active and did not altogether acquiesce in the care given her by her daughters. She was very
proud of her old things, especially her dishes and china which she
delighted to display of an afternoon when relatives and old friends
called for a cup of tea. The cookies and cakes were of her own baking—
she could not trust even her own daughter on such an important matter—no one could make them to suit her.

One uncle had a good sized flour mill near a village called Doon. This was my especial delight. The over shot water wheels and the five grinding mills of huge mill stones and numerous belted bolting and separating machines and the cup conveyors to the various floors, were a liberal education to a mechanically minded boy.

All but two of my uncles lived on farms. Besides the miller, one had a large button factory in nearby Berlin (now Kitchener). As he had travelled widely in South America and Europe, he was the wonderman of the community to whom all stood in awe.

These Pennsylvania Dutch or Swiss neighbors married neighbors. Slichters married Hubers, Detweilers or Hallmans, and Hubers and the others did the same. There were scores of double cousins among them. The parents were land minded and always ready to help buy a farm for a son. They found all the good life they needed in work, worship and song. Perhaps they were wiser than they knew. Their troubles were only the troubles of living, the inescapable troubles common to all; and for these they had at hand the ever-present help of their God, in whom they confided and communed and to whom they consecrated their daily labors. The well worn Bible was a book of knowledge but also a book of poetry and drama, and when age numbed their senses and their strength, the good book did not let them down.

But that life of security was under attack. Industrialization was closing in with crushing speed upon that simple society. Youth began to forsake farm for city. The spiritual and physical security of the contented but arduous life on the land was bartered for the spiritual and physical insecurity of the city. An epoch, an old and tested form of living was moving into the shadow. Few were those who were equipped to meet the new and untested mode of life. There were no family groups that could kneel in prayer before beginning the day's work in factory or office. One could not take his zither to his job in the mill. There were no Swiss songs that could be sung to the rhythm of shop and counting room. More than all, the old guides of religion did not seem to be needed or to be of help. It was easy to forget to call upon one's God for aid when it was a machine and not a man that did the work.

Schooling. We know little of the precollege schooling of Charles. We do know that, during the last two years before going to Northwestern University, he attended the Oakland High School where he evidently made a good record. In 1911 he discovered that his high school principal, C. I. Parker, was still active and, on April 4, wrote him:

~ My dear Professor Parker:

I was delighted to receive direct word concerning you from our Professor Hart. It has been many years since I have seen you personally, and it delighted me to know that you are well and vigorous.

I still keep at work at about the same things that I worked at thirty and thirty-five years ago, when you were principal of the Oakland High School. My family consists of four boys, two of them in the Madison High School, and another is a Sophomore in the University. The fourth just finishes the Grammar grades. These boys are just back from a year of study in Germany where I expect to send them again in two years. They seem to be industrious fellows, and I believe will have no difficulty in excelling their father in their life's work.

It can do no harm to again express my appreciation of the great help you have been to me especially in my early days when a good start meant so much for the future. I shall never forget the most cordial letter you wrote for me when I left for college, after two years spent in your high school. I assure [you] that I have kept your letter to the present day and have constantly endeavored to live up to the expectations you so cordially expressed at that time. With best wishes, please regard me

Most affectionately yours,

The reply from Parker was equally warm.

The first literary attempt that we have from the boy's hand is dated February 25, 1881, when he was a senior in high school, and was evidently something he enjoyed since he kept it for the rest of his life.

A certain Mother Shipman prophesied that the world would come to an end by fire or, as Charles wrote, in a "final confligation" in 1881; and a classmate of his, Mamie Hood, must have taken seriously this forecast, for young Slichter produced (probably for a high school literary club) a reply to Mamie's essay in the form of "Karl Schlichtehr's Prophecies." Karl was claimed by Charles as an intimate friend and bedfellow. The buildup is typical of the high school humor of a very bright boy and the prophecies are purposely banal. The first was: "Prophecy the first:—Be it known that the Fourth-of-July will come on the fourth day of the seventh month of the year, 1881"; and the last was: "Prophecy the sixth:—Be it known that there will be many people who will be fools enough to believe Mother Shipman's prophecy."

Perhaps he could have become an adept muckraker for, in April 1881, he wrote a paper entitled "Can a Businessman be a Christian?" (I suppose, however, that a professional reformer must have more fervor than humor, and that was seldom true of Slichter.) He gave a

lively account of a Chicago wholesale firm of three brothers: the poli-
tician, the playboy, and the religious moralist. Any customer could be
fitted for style. But even the churchgoer, though shocked at the suggestion of a business engagement on the Sabbath, found the idea less
shocking than that of losing a several-thousand-dollar order.

A quarter of a century before the Pure Food and Drug Act, this same
paper contained a description of how vinegar was made from sulphuric
acid combined with decaying vegetables, and it continued:

~ But there is a greater deceit in this "Genuine Horseradish" which
is bottled for table use. It is a mystery how they can grind and sell
horse-radish during the whole year when the genuine root is only fit
to be put up at a certain stage in its growth. The secret is this: while
dutch turnips are being ground to sawdust in a suitable a solution of
Aqua Ammonia is added which converts the turnips into the most
salable Horseradish conceivable.

The teacher's comment was: "Charlie, why don't you answer the
question you have asked? Cannot a successful business man be honest?"
which I take was not a criticism of the paper but a challenge to the lad
for his future. The teacher also indicated the above paragraph needed
some reconstruction.

Northwestern University

From the time Slichter went to college in 1881 until his death in
1946, we have much evidence of his activities. Concerning his four
years at Northwestern there are three types of documents: official
records; themes which he saved (justifiably, he was one of his own
favorite authors); and anecdotal material, much of it recollections on
his part or on that of others of days long past. Fortunately, even if
memory with time becomes less exact, it does not become less colorful.

Slichter's academic record was excellent. His grades ranged from 83
in International Law to 97, once in Chemistry and twice in Composition; 98 in Natural Theology; 99 in Geology; and 100 in Logic, Zoology, and two times in Biology. When he graduated he was one of
seven students to get "General Honors—First Grade"; and he also got
"Special Honors" in Natural History, Mathematics, and Physics. He
was the only student in his class to get "Special Honors" in three
subjects. After the chapter of Phi Beta Kappa (alpha of Illinois) was
established at Northwestern in 1889, he was elected as an alumnus.

His bachelor's degree was not the last degree he was to receive from
Northwestern, for in 1888 he was awarded a master's degree in science

under conditions described in Chapter 6, and in 1916 he was given the honorary degree of Doctor of Science, as befitted one whose special graduating honors were in three science departments. This must have been especially satisfying to him, since his escort was the affectionately admired Dean Thomas F. Holgate.

There is reason to doubt that his deportment was as exemplary as his scholarship. Twice in the spring of his senior year the secretary of the faculty asked him to explain to the president in person a series of unexcused absences—nineteen in all—only one in science, but four in International Law, four in Christian Evidences, and six absences from Prayers.

There may have been more active misdemeanors for, on June 12, 1905, Rush McNair, a classmate, wrote to Slichter explaining that he could not come to the class's thirtieth reunion and remarking:

~ *There was a time after old Joe found your britches in the stairway of the clock tower that it may not have been wise to have been caught in your society.*

But now I would not hesitate so to be seen; in fact I would delight in the honor.

In later years Slichter advocated that everyone learn to play poker, a game for which he claimed great educational value. He would explain that, since poker was frowned on by the authorities of Northwestern University, he learned to play in the room of a student in the theological seminary, a region beyond official suspicion. The student who furnished the sanctuary became a bishop and, as Slichter proclaimed, "a damn sight better bishop for playing poker."

His unfrowned-on interests seemed to have been quite normal. He belonged to Sigma Chi, engaged in securing advertising for student publications, and had, for the consideration of one dollar, a pass to the baseball games. These activities seemed neither to have furthered nor hindered his later distinction.

One activity which probably was also connected with his high school days, must have won the hearty approval of the administration: In 1896, when urging that the University of Wisconsin recruit students from Chicago (". . . A place where there is hunting, fishing, sailing, iceboating, etc., has an attraction for a Chicago boy which it is difficult for you to realize."), he stated: "Beginning in 1882 I furnished to Pres. Cummings of Northwestern names and addresses of all students in Chicago High Schools. These I could obtain gratis, being connected with the Cook County High School Journal."

Slichter's undergraduate writings are of considerable interest, for they indicate his developing literary skill as well as his ideas, at least one of which was to be a leitmotiv; throughout adult life he championed youth—first apologetically and finally with strong conviction.

In the winter of 1884 he turned in a paper entitled "The Young Man—Mostly in his Defense." In spite of some wild rhetoric, the paper deserved the 95 which it received. Alongside the sentence, "A perpetual stream of impossible logic has managed to supply the popular discussion with fire and smoky doubt," the instructor wrote, "A stream does not commonly supply fire." This paper also has some reflections typical of Slichter's maturity.

~ *People are not born with characters ready formed—they must be made for them by a system of training and correction. The boy who needs correction and gets it has in every way the advantage of him who has never needed it. There is nothing permanent about school-boy tricks; but, on the contrary, the time will inevitably come to you when the recollection of them will set you thinking. The remembrance of your puerile rascality may frequently be a great help to you, while the memory of what a cherub you have been will only lead you to fatal presumption.*

Do not despair of the young man as long as his disposition is not cowardly. Whatever his character, as long as there is bravery in it, there is no limit to the possibilities to which it may develop.

His philosophy had not changed drastically when sixty-one years later he wrote, as class president, to his six surviving classmates living in five different states:

~ *Dear Eighty-Five:*
This is the sixtieth anniversary of the graduation of the famous class of 1885. We should have a reunion, of course, but the O.D.T. says it should not involve travel, hence we must reunite by letter. That is lucky, for the seven members of 1885 are widely scattered and thousands of miles off the center from where we separated just sixty years ago.
Time makes some shifts and some readjustments in judgments. Those gaudy caps we wore as freshmen, and those "mortar boards" we wore as sophomores, and the gray "plugs" we sported as juniors, and the silken tall hats we flaunted as seniors, do not seem quite true history now. Perhaps all this meant that we were bound to be different and just a little on the youthful and funny side of life. But whatever our indiscretions, they did not stop us from growing up. Our youthful foibles meant that we needed more time to mature. I am glad that we

started out so green. I resent the fact that the world has tried to ripen and yellow us so fast. It is best to stay green and not stiffen with the tough bark and dried out pith of age.

That Trig funeral and the mottoes painted on the Chapel wall are soon to be the forgotten part of history—it may be fortunate if these stories are missed by our grandchildren.

With memory and love to all 85, I remain,

In his old age Slichter would have defended "The Young Man," but he would have found much of the doctrine of his sophomore composition, "The Tendencies of Modern Science" foreign to his thought. No one would guess that Slichter had written: "The tendency of Scientists to go beyond the bounds of their own department is a most disastrous one. The Physicist wanders into the domain of physiology, the physiologist into the limits of metaphysics and the metaphysician into regions in which no Scientist has any business."

In spite of this sophomore dictum he was already engaged in bringing one science to the service of another, for, as described in Chapter 4, his first published scientific paper, an application of physics to biology, appeared in 1884 while he was still an undergraduate. In "Tendencies of Modern Science" he wrote:

~ The theory of evolution furnishes another impressive example. This doctrine, proportioned and ornamented by its dogmatic teachers, has been received by hosts of hot enthusiasts as an invincible article of faith. But how many of its disciples can state the foundations for their belief? How many have ever thought of testing the assertions of its exponents? Far from being established, there seems but little prospect of its ever becoming anything but a doubtful hypothesis. To say that mind is the creation of matter, is materialism in its worst form, and something which lacks proof.

However, in 1904, there was an exchange of letters with his brother Frederic. On October 19 Fred wrote to Charley:

~ Please drop me a line to my home address, . . . what later authorities "intelligent people" follow on the evolution of man. In my library I have Huxley, Spencer, Muller [?], Darwin, Redpath and others but the rector of our Church, who is also archdeacon tells Anna that "intelligent people" now do not place any credence in the theories advanced by such writers as above referred to. If there is any later theory I wish to buy the books and keep up to date.

To which he received the reply, dated October 21:

~ I have your favor of October 19th referring to the present condition of the Theory of Evolution advanced by Darwin and others. This is a line of thought on which I do not feel myself as especially competent to speak authoritatively, but I believe there is no doubt but that the fundamental doctrines are still considered, by scientific men, to be essentially correct. Like all other scientific truths, the work of [a] multitude of scientific men has been able to add to and augment the first statements by Darwin; this does not mean, however, that the science expounded by Darwin was not true, any more than the great additions that have been made to the Science of Mechanics since the days of Newton are to be understood as substituting the fundamental truths conceived by Newton.

Fred again wrote on October 25, a most human letter:

~ Dear Charley:
    Yours 21st rec'd. As I have read many thousand pages on the subject, besides about everything of authority on ancient history that [I] could buy, [I] believe my own opinion quite as valuable as that of our rector who probably likes to believe the scriptural saying: "And God made man in the image of Himself, and he was perfect." Yet when you have a family of children the father's opinions are occasionally discounted. I prefer the girls to read and then form their own opinions and such letters as yours assist materially in inducing them to do so.
Of course we have advanced as a race since the days of Alexander or Xenophon, just as they were far superior in many ways to the cave dwellers. At any rate, we don't aim to eat any more bugs or worms than possible.

The year after Slichter graduated from Northwestern he taught at the Athenaeum in Chicago. We have no account of this experience. In 1886 he joined the teaching staff of the University of Wisconsin.

c১ *Department Member*

IN 1885 the Department of Mathematics at the University of Wisconsin consisted of "Professor Van Velzer, assisted by Mrs. Carson," and in 1886 "Professor Van Velzer, assisted by Mr. Slichter." Until 1892 these two were the only full-time members of the department. In 1889 Slichter became an assistant professor, and in 1892 was promoted to a professorship. (This latter promotion is one of the earliest evidences we have of President Chamberlin's respect for Slichter.)

The letter that Van Velzer wrote Slichter on July 3, 1886, follows:

~ *There is a temporary vacancy here in mathematics which I have
already offered to a young man but have not heard from him yet and
I am afraid he does not want the position and as I am in something
of a hurry to fill the vacancy I write to ascertain if you would like the
position in case the other young man does not want it. The position is
that of Instructor in mathematics and is a temporary one for one year
at a salary of $600. There is some probability that my assistant (who is
away on a leave of absence for one year) will not return to her work
in mathematics in which case there will be a permanent position here
that of Assistant Professor of Mathematics and in that case the person
appointed now, other things being equal, will stand the best chance of
being elected to that position. If you would like the position under all
the circumstances please answer by return mail.*

*The salary attached to the Assistant Professorship will probably be
$1500, but it will take 3 or 4 yrs. to work the salary gradually up to that
point.*

(A third of a century later another professor of Dutch extraction made even clearer to me that my first appointment at the University of Wisconsin was for one year only.)

Charles A. Van Velzer came to Wisconsin in 1881 to assist Professor John W. Sterling, who, on February 5, 1849, had taught the first class at the University. Van Velzer remained until 1906 when, being faced with a choice of resigning or giving up a business that Dean Birge and others believed was taking an exorbitant amount of his attention, he resigned. He later taught at Carthage College to an advanced age.

Teacher

Cold Facts.   We start with a few statistics.

David Allmendinger made an analysis of the instructional reports of Professor Slichter. Being a historian and hence more accurate than a mathematician, Allmendinger made some corrections and raised some questions. The following figures are therefore subject to minor alterations but in general give a true picture. Slichter taught over 30 different courses, some of which overlapped in content. Most were either elementary or in the area of applied mathematics. Those that he taught more than 10 times follow: (The figures after the course titles are the number of sections of the subjects reported by Allmendinger. These are minimal, for some of the courses probably were given also in terms where the records are missing.) Theory of Elasticity—10, Theory of Equations—10, Theory of Probabilities—17, Thesis—19, Hydrodynamics—19, Theory of Potential—23, Analytic Geometry—27, Trigonometry—30, Algebra—40, Calculus—48, and Theoretical Mechanics —79 (almost an obsession), first given by him in 1894 and last in 1933.

Perhaps even more significant is the large number of advanced courses he taught only a few times, for they indicate a desire to learn as well as to teach. Such courses included: Dynamics of a Particle, Fourier Series, Theory of Attraction, Newtonian Potential Functions, Spherical Harmonics, and Differential Equations. In spite of the fact that his formal mathematical education was meager, Slichter acquired a working knowledge of applied mathematics which was equaled by few of his American contemporaries.

Not counting summer schools and a few terms for which the reports are missing, and recognizing that there were many who took more than one course with him, the grand total enrollment in Slichter's classes was 6,421. (I would guess well over 3,000 different persons.)

It is interesting to divide the distribution into four periods: the decades 1886 to 1896, 1896 to 1906; the period of his chairmanship, 1906 to 1920; and of his deanship, 1920 to 1934. Distribution among the classifications and the totals are as follows:

APPROXIMATE ACADEMIC YEAR ENROLLMENT IN CLASSES OF C. S. SLICHTER

Classification	1886–1896	1896–1906	1906–1920	1920–1934	Total	Percent
Graduates	25	164	343	214	746	12
Upperclassmen	102	75	244	126	547	8
Sophomores	535	510	546	4	1,595	25
Freshmen	1,831	889	260	0	2,980	46
Special	428	89	1	0	518	8
Undetermined	31	0	1	3	35	1
Total	2,952	1,727	1,395	347	6,421	

An explanation of the classification "Special" is given in the catalog of 1887–88, the year after Slichter joined the University, as: "Candidates who do not desire to graduate or who wish to select their studies, and those who wish ultimately to obtain a standing in some regular course to which their preparation is ill-adjusted, are permitted to take special selected courses. Such students may enter at any time and take any studies which they are prepared to prosecute to advantage." The same catalog lists 637 students, of whom 146 were "special" students. There were four "resident graduates." To have been told that some day he would be graduate dean scarcely would have seemed exciting to Slichter at the time.

The fact that nearly half of his class enrollments appeared in the first decade results in part from the University's academic year being divided into three terms rather than two semesters during nine years of the decade. After he was promoted and Skinner joined the staff, Slichter more regularly taught calculus, which caused a shift from freshman to sophomore enrollment in the second decade. The increase in his graduate instruction of course reflected a university-wide phenomenon, but it also reflected the fact that during Slichter's earlier years Van Velzer had taught much of the advanced work.

In the first decade he never taught less than fifteen hours per week, and frequently as many as twenty. The load decreased in the second decade, ranging from twelve to nineteen; as chairman, it ranged from twelve to fifteen; and as dean, averaged a little over three-and-a-half hours per week.

In addition to the work carried on in class, three Ph.D. theses were written under his direction: Henry C. Wolff (1908), "The Continuous Plane Motion of a Liquid Bounded by Two Right Lines"; Thornton C. Fry (1920), "The Use of Divergent Integrals in the Solution of Differential Equations"; and William E. Cederberg (1922), whose later ca-

reer was at Augustana College, "On the Motion of a Double Pendu-
lum."

Warm Facts. It is impossible for me to write this section without
being somewhat autobiographical, for I was one of Slichter's students
in the Theory of Probability. I shall compare him with two other great
teachers of mathematics whom I have had: Professors E. H. Moore of
Chicago and E. B. Van Vleck. Slichter had less influence than either
of these two on the development of mathematics, but his teaching
probably had more influence than theirs on a great body of nonmathe-
maticians, for Slichter was a splendid teacher, especially of students
who cared more for the use of mathematics than for its development.

E. H. Moore did research at the blackboard. It was pure excitement
for one of mathematical talent to be in his classes, but it was frustrating
beyond measure for others. He labored with those who wrote theses
under him. Technically he was a poor lecturer, and I judge that he
was not a successful teacher of undergraduates. Many years ago R. G. D.
Richardson, then secretary of the American Mathematical Society,
said to me: "E. H. Moore is the worst lecturer I have ever heard"; and
a few minutes later: "Moore is the greatest teacher of mathematics
America has ever had." For decades American mathematics was led by
his students, grandstudents, and great-grandstudents.

Van Vleck, although at times his tongue stumbled over itself trying
to catch up with his brain, gave finely organized lectures. He knew
mathematics was an art and he respected its beauty. His proofs were
both elegant and rigorous; so were his writings. Unlike Moore, he seldom attempted to lead his class into the unknown, for he was a lonely
knocker at its gates; but he did present mind-stretching material in
polished form. Outside of class he gave himself generously to his students. Like Moore, his teaching chiefly influenced mathematicians.

In class, Moore in seeking new mathematics and Van Vleck in presenting it, although never forgetting the quest or the exposition, could
become so engrossed in the subject as to almost forget the student. Not
so Slichter. In his wondrous address, "The Self-Training of a Teacher,"
delivered September 5, 1931, he said:

~ *Forty-five years ago this very day—almost this very hour—I began
my university teaching at Wisconsin. I cannot think of anyone less
prepared in technical psychology and pedagogy for such a position. Experimental psychology was then just beginning, and the science of
education was groping about like a small child. Perhaps, therefore, I
should also say that I know of no one who at that date was better pre-*

pared in technical pedagogy for university instruction. I started off with four daily classes of forty freshmen each and an advanced class of two students, which soon dwindled to one. The universities of America have always been generous, even overgenerous, in supplying an abundance of students to their green and utterly inexperienced tutors. I was so unsophisticated that I could think of no finer job than to teach mathematics to freshmen. Until quite recently I thought that everybody was of the same opinion. It was a shock to me to learn that some think that there is a higher job.

But actually I did not teach freshmen. I taught attorneys, bankers, big business men, physicians, surgeons, judges, congressmen, governors, writers, editors, poets, inventors, great engineers, corporation presidents, railroad presidents, scientists, professors, deans, regents, and university presidents. For that is what those freshmen are now, and of course they were the same persons then.

I was teaching with such enthusiasm and was so proud of my classes that it never occurred to me that my work could be criticized. It was the self-confidence and conceit of a beginner. But President Bascom visited my classes. After each visit I was called to his office. He went over things with me criticizing at length, but justly and with uncanny precision and directness, calling upon his imagination to say something pleasant at the end. Why did President Bascom take such trouble with an insignificant tutor? It was because he knew that I was not teaching freshmen, but the men I have mentioned. That is the first lesson that the university teacher must learn. Our contact with students must not be for a moment nor at a point, but through the complete trend of a life, the years to come being of one substance with the present. We must be ready by our magic to unroll the long scroll of a life before us to the epoch of the present or to the epoch of the future, as we will. A prophet once exclaimed: "Behold the dawn! Behold the dawn! The things as yet but half declared command the coming day."

It is perhaps worth noting that in his list of those he taught he omitted mathematicians except to the degree that they are included among the poets and scientists.

It is also in this address that he said:

~ The third lesson for the college teacher to learn is the art of making his subject interesting. It is not his task to teach interesting things, as the quacks proclaim, but to make interesting the things that must be taught. He must teach the basal sciences as they exist, and it is his task to make the things they contain a part of a living world, full of the adventure and the hazards of learning. This, of course, is the reason why an instructor must be highly trained and productive in his special field and widely read in the history of science, and also have some knowledge of the literature of the race.

This last point deserves further elaboration, especially for the younger instructors. It is a sad experience to visit a classroom and discover many rich personalities on the students' side of the rostrum but a weak or undeveloped personality on the teacher's side. It is obvious that the teacher will find great difficulty in overcoming this potential difference. How may the teacher develop a rich personality? Unfortunately there are not psychological calisthenics that can transform a weak into a rich personality. It is largely a matter of inheritance. Luckily, however, it is not a matter of family but of racial inheritance. We are all mentioned in the wills of Homer and Shakespeare and of all the great masters of letters; we do not share merely in part, but each of us inherits in full all their rich chattels. You may be assured that great literature, great history, and great biography belong to us—even to mathematicians—and may be claimed and cultivated as our own. Hence the students have a right to find something more in the classroom than the narrow mechanics of a scientific machine. Instead of twenty times at the movies, better the guest of Homer twenty times. What has Homer to do, you will ask, with the teaching of mathematics? I use Homer, of course, merely as an example. His heavy-breasted heroes contending in the sweat of their primal passions are reacting to the same emotions that sway men everywhere, and to motives that rule our own destiny. As you follow the story of their strife the fibers of your being will vibrate with the ambitions and angers universal among humanity; your personality will quicken to a new appreciation of the facts of life. In other words, the ingredients necessary to compound a man can hardly be omitted from the recipe for making a teacher. You need not be afraid, therefore, to enrich your life by the cultivation of letters and by indulgence at the feast of the humanities.

As you face the Memorial Library of the University of Wisconsin you may see engraved to the left of the door:

> BOOKS ARE THE SEED
> WHICH IS RETURNED TO THE SOIL
> AS THE CONDITION OF
> FURTHER INCREASE.
>   JOHN BASCOM

And to the right:

WE ARE ALL MENTIONED
IN THE WILLS OF
HOMER AND SHAKESPEARE.
CHARLES S. SLICHTER

Four years before Slichter gave the above-quoted address, "The Self-Training of a Teacher," he replied to Dean Otis E. Randall, of Brown University, who had suggested that graduate students preparing for college careers take more work in the methodology of teaching:

~ *The proposal implied by these questions undertakes to continue our present absurd plan of trying to make college teachers and scientists out of incompetents. We have a group of graduate students that seek to enter the higher positions in the secondary schools and colleges who lack personality and virility and seem to desire to take refuge in the teaching profession because they lack the courage to face the chances and hardships of a business or professional career. . . . You seem to plan to make men and women by courses in pedagogy out of such material. It is hopeless. The real problem is to divert from the stream now entering the world of business enough of the able fellows to enter the profession of teaching.*

I vividly remember the course in Probability which I had under Slichter. The major theorems were developed and their uses explained. The proofs were often antiquated, sometimes even tedious, but the discussions of the applications were magnificently illuminating: Gauss and the Theory of Errors and Least Squares; Galton, exploring the relation of the statures of fathers and sons through the Regression Equation, incidentally showing why the equation received that name; and Bayes' Theorem with the remark, "With this theorem you can tell which neighbor's bull jumped the fence." I never got this kind of insight into the uses of mathematics from any other teacher, but, on the other hand, I have seldom been exposed to a more cavalier treatment of mathematics itself.

Now I must unburden myself of a grudge I have kept hidden for nearly two-score years, a grudge against both Slichter and Van Vleck. After Slichter had given an outrageously awkward development by means of Sterling's approximation formula of the normal probability distribution, I worked out what I thought (and still do) a far more elegant approach. I showed it to Slichter but he seemed uninterested. The theorems were already known. Why seek a new proof? New uses would be more to the point. I had better show it to Van Vleck who would be more able to criticize it. I did, and Van Vleck merely remarked that it lacked a good deal in rigor, but he did not acknowledge its basic elegance nor suggest how to improve it. Anyway, I have used it ever since!

The above paragraphs could lead one to believe that Slichter did not appreciate mathematics as such. This would be far from the truth. In a carefully prepared paper entitled, "Discussion—Joint Meeting of Engineers and Mathematicians," undated but probably given about 1912, he stated:

~ If we teach a trade and not a science the time is largely wasted. If we teach dyeing and not chemistry, the graduate is already out of date when he begins his career, and he has not the fundamental principles wherewith to bring himself abreast of the times. I therefore regard it of greatest importance that mathematics be taught to engineering students with real enthusiasm for the science itself. It should be taught by men who themselves are actively contributing to the growth of mathematical science. The present spirit of engineering science is such that no instructor in any of the basal sciences is satisfactory who does not see that it is his duty not only to teach what is known, but to be interested in and to take an active part in the development of what is unknown.

After describing the engineers he had hired in connection with his work for the United States Geological Survey, he continued:

~ I find it true that the boys have forgotten a great deal of the material they had in college, and that they have remembered other things. They remember the manual and the mechanical things—how to swim, how to ride a horse, how to fish, how to play ball, how to run the level, how to work the plane table, and how to do stadia work. Now what have they forgotten? The men have forgotten the intellectual things—hydraulics, electrical science, thermodynamics, etc. The human mind possesses an unlimited capacity for forgetting. But my experience shows that the young men forget their hydraulics just as much as they forget their mathematics or their mechanics. The engineer in the field observes that a boy remembers the right end of an instrument and seems to be amazed that the same man does not know the right end of an integral sign. He therefore concludes that the mathematics has not been "taught right." If he will compare intellectual things with intellectual things he will find that a miscellaneous group of engineers will pass as good an examination in mathematics ten years after graduation as they would pass in thermodynamics or hydraulics.

It grates on me to hear mathematics spoken of as a tool. Mathematics is to the engineer a basal science and not a tool. The spirit of that science is of more value to the engineer than the particular things that can be accomplished. The engineer need not be a mathematician, but he needs to think mathematically, and, to my mind, he needs the power of mathematical thought more than skill in manipulating a few tools in mechanical fashion. There are already too many factory made products turned over to the college by the primary and secondary schools. I make a fundamental contrast between the engineer with his mind endowed with the power of creative and rational design, and the artisan with his hands equipped with tools for physical construction. A great engineer must be trained in correct seeing and thinking and must have the power of reasoning concerning some of the highest abstractions of the human mind. In this aspect mathematics is not a tool—it is a basal science.

This represents Slichter truly. But the portions of mathematics he cared most about were on the boundaries with other sciences, for to him the context of mathematics was as important as its structure. Moreover, he was not entirely joking when, at the dinner of the 1927 summer meeting of the American Mathematical Society at Madison, he acted as toastmaster and declared: "All science is mathematics. Physics is the noisy part of mathematics, chemistry is the smelly part of mathematics and biology is the messy part of mathematics."

It was on this same occasion when, in introducing the president of the University, he said, "And now while he is speaking, ponder the question: Is time infinite, or does it only seem so?"

The first student comment on Slichter as a teacher which I have found, is an entry in the diary of Sidney D. Townley, a member of the University of Wisconsin class of 1890. At the time of writing, October 10, 1886, he was a freshman; he was to become an eminent astronomer. During 1939–40 he had a portion of his diary reproduced under the title of *Diary of a Student*—for distribution, I presume, chiefly to classmates to help celebrate the fiftieth anniversary of their graduation.

~ Am very busy during the week with my studies. We, general science students, part of them, recite in German from 8 to 9 each morning to Julius Emil Olson, B.L., who also teaches Scandinavian languages. He is a good teacher. From 11 to 12 Algebra to C. L. Slichter, B.S. He is a nice young fellow and a good mathematician. From 12 to 1 on Monday, Wednesday, and Friday a lecture on botany from F. L. Sargeant. He is a queer man, comes from Harvard, and is a good botanist. From 2 to 3 on Tuesday and Thursday a lecture from Edward A. Birge, Ph.D., on zoology. He is a queer looking man, but everyone that has anything to do with him likes him. He is a very good teacher. From 9–11 on Tuesday, Wednesday, and Thursday I have laboratory work in Biology. For the first two weeks we studied microscopic animals and by means of our compound microscopes were enabled to study animals *of only one cell, with neither legs nor wings and no opening in its body, yet as truly an animal as man himself. We are now studying the structure of the lower orders of the vegetable kingdom. We are obliged to make drawings of all things examined. Biology I think is a very interesting and useful study. From 3:30 to 4:15 on Tuesday elocution by Fred J. Turner, A.B. At same hour on Thursday instruction in composition from W. S. Tupper, A.B. I am required to have an essay and two declamations this term. Last and least, drill from 4:30 to 5:30 on Monday, Tuesday, and Thursday, and from 4:30 to 6 on Wednesday and Friday by Luigi Lomia, M.S., 1st Lieut. 5th U.S. Artillery. We all dress in blue uniforms and caps, white gloves and black cravat and white belts on inspection day, which comes every Friday. We have been drilling so far in squads of eight or nine each; next week we will be organized into a company and commence drilling with guns.*

Nowadays, freshmen, or at least their parents, complain of how few professors they have in their classes. In 1886, Olson, Slichter, Sargent (correct spelling), Turner, and Tupper were "Instructors or Assistants," a single category in the catalog of that year. Olson, Turner, and Slichter remained for many years at the University. The first two have dormitory units named after them and the third, a whole building; also there are professorships carrying the names of Turner and Slichter. Sargent became a noted botanist and author. If freshmen fare as well today, they should not complain. The assistants today are better prepared in their fields than those in 1886. It will take half a century to tell whether they were as well selected.

In 1943 another alumnus reminded Slichter of the year 1886. "You know we entered the University on the same day. I was learning to square 'A plus B' and you were learning to look dignified."

In midcareer, we have Professor Otto Kowalke's recollection, given to Mr. Allmendinger, of Slichter as a teacher. Kowalke was for many years a Professor of Chemical Engineering and was himself a beloved teacher. We follow, mostly in his own words, Allmendinger's notes of January 11, 1966:

Kowalke remembers Slichter as one of his best teachers, though he had him for just this one course. Kowalke enjoyed the algebra (and trigonometry and bits of calculus) he learned that semester. "Oh yes, it was fun," he says, "I enjoyed that." He learned a lot, for Slichter expected a lot of work, but Kowalke would not use the word "demanding" to describe Slichter's manner. There was never the sense that you would be rapped on the knuckles if you didn't perform. Slichter was inspiring, and a leader. The boss. He got the boys to work by making them feel they ought to perform. He says Slichter had no special methods or techniques that he could think of; he accomplished everything through his personality. Slichter made no references to his own research in the course. (Understandable: Algebra.) He made no special demands and seemed to admire no special kind of student. He simply wanted to inspire all the boys. Kowalke says he didn't ever fear Slichter; he just followed his leadership.

Presumably there would be those today who would criticize Slichter for spending a recitation period on current events rather than mathematics. That is what he did the morning after the Battle of Manila Bay. In 1943 a member of the class of 1901 recalled the event and wrote: "Your talk that day—it took up the entire hour—was one of those inspiring hours we do not forget."

And again in 1943, another student declared that he had learned more from seeing Slichter "the great mathematician" praying at a midnight Christmas Eve service than from all the lectures he had heard at the University.

The latest written comment (oral comments are continuous) on Slichter's teaching which I have received, was in a letter from Thornton C. Fry, dated September 17, 1969. Fry, as already mentioned, was one of Slichter's three Ph.D. students. He had served as assistant and instructor at the University, and for many years was with the Bell Telephone Laboratories and its predecessor—the research branch of Western Electric. He lectured frequently at various universities and has been an active leader both in the American Mathematical Society and the Mathematical Association of America. Fry's thesis subject, "The Use of Divergent Integrals in the Solution of Differential Equations," did not sound like one that would be expected of a dissertation written under Slichter. Therefore I asked Fry for the facts. Part of his reply follows:

~ Your remarks about my thesis are very perceptive. It grew out of my personal interest in putting a foundation under some formal results of Heaviside, which were a lively topic in communication circles at the time. I was then at Western Electric (Bell Telephone Laboratories had not been formally created) and John Carson was at AT&T. We approached the matter from different angles, and both succeeded. I wrote a paper for publication and sent it off before the idea occurred to me that perhaps with a little more effort it might become a thesis. When I got the idea, I sent a copy of the paper off to Slichter to get his reaction. About a week later I got a reply of eight or ten lines, as I remember it—my records are now in storage, so I cannot look up the original—which said in essence, "Your thesis is accepted, come out and get your degree." Slichter had not seen the document before, and no changes were ever made in it afterward. He was, of course a member of my examining committee (along with March, Comstock, Mason and Ingersoll—a pretty distinguished group!) but as far as I can now recall, neither he nor any of them questioned me about my thesis. I think I must have been a complete maverick.

This, however, does not mean that he had no influence over me. He not only hired me as an assistant instructor and helped solve my financial problems in other ways; he also did as much or more to convert an unusually immature kid into a scientist as anyone else I came in contact with. (The others at Wisconsin that influenced me most were Comstock, March and Dresden.)

He was an unusual person, in appearance, in personality, and as a teacher; in all respects there was something charmingly outlandish about him. He had a dark Bedouin complexion, piercing mesmeric eyes, and thick gray hair; it was impossible not to notice him either in a gathering or when he was walking along with his peculiar slow swinging gait. He was gruff, witty, demanding and infinitely compassionate. The independence of his methodology as a teacher is well illustrated by the course on Elementary Analysis which he set up for engineers; and his personal eccentricity by his fixed habit of adopting nicknames —Bill, Joe, Jim, etc.—for his students, undergraduate and graduate alike; as well as his frequent habit of putting one foot in the chalk trough while lecturing, and sometimes writing left handed over his shoulder while it was there.

He should have been more careful, for some fifty years later this contortionist act came near to crippling two of the best, but not youngest, secretaries in the University who tried to test the story by the experimental method. It did not occur to them that Slichter may have been seated.

The Legend.   The stories about Slichter's teaching are legion. A few may be inventions; a few may be the attributing to him of tales told of many teachers; a few may be the accretion of episodes that happened at Wisconsin and, in the retelling, were assigned to the most likely candidate; but most I believe to be fact or the elaboration of fact. The elaboration of fact would not have been distasteful to him. Although the following legends have not passed through the sieve of biblical criticism, they depict the man—even the few that may not have happened. The fables that cluster about an interesting person belong to his portrait. In some instances the sources are indicated, but in other cases they are forgotten by me or by those who related them to me.

Some of the best anecdotes arose from the tendency of Slichter to come late to class, a habit that he seems to have acquired early and, as I know, retained long. When he found the class shying their hats toward the chandelier and as one successfully lit upon it, he exclaimed: "That's the brightest thing that hat's ever been on." Another time he was quite late and all the students except two had left. When he arrived the conversation was as follows: "Where is the class?" "They have left, sir." "Why?" "Because you are late." "Why are you here?" "We thought we ought to stay, sir." "Get out! If you haven't learned to cooperate with your fellowmen by now, you will never be a success."

Another version (not really incompatible with the above) is that, as he came up the hill, he met some of his class and, being informed of the too faithful two, said: "Well, let the darn fools wait." He may have gone on to his tryst with the "darn fools."

Professor Kowalke tells the following: The class, which met in a room on the ground floor, appointed him their sentinel one day. Slichter's usual course was to come down the stairway in the President's offices. Kowalke waited at the bottom of these stairs, while the boys in the class waited for his whistle. When he heard Slichter coming—on the run—Kowalke whistled and ran back to the classroom. He was the last one out the window; Slichter caught him by the leg and pulled him back down into the room. They stood facing one another, Slichter laughing very hard. The next day Slichter appointed another student as timekeeper, telling him to allow him three minutes and then to dismiss the class.

Many, besides Fry, have told of the liberties he took with students' names. My wife's brother-in-law, Howard Beasley, would be called upon as "Hey, you, Beeswax."

On January 12, 1934, at the dinner given for Professors Slichter, Skinner, and Van Vleck, Warren Weaver described how he fared in this regard:

~ Twenty-one years ago last fall there walked into one of the basement rooms of what was then Main Hall, a green, timid, and flap-eared freshman. At the front of the room was a man with a deep chest, a thatch of iron-grey hair, and a booming voice that caught that scared freshman in an explosion of vocal energy and said: "Bill, you'd better sit down and stay awhile."

The freshman did stay for nearly twenty years, and for well over a quarter of that interval he sat, semester after semester, in this man's classes and was boomed at, continuing to be Bill except for one semester when, for an entirely mysterious reason, he spent that period disguised as Jim. All this time the freshman's wonder and respect and admiration grew; but never, let me assure you, would he have dared to turn the tables and do a little booming himself.

It is also Weaver who in his autobiography, *Scene of Change*, writes as follows:

~ *I must admit that he [Slichter] both confused and frightened some of the poorer students. I recall an earnest boy who asked the professor a stupid question. Slichter paused for several ominous seconds, and then with unexpected calm said, "Bill, suppose you stay after class and ask the janitor."*

One of his secretaries told Mr. Fred that if Slichter wanted one of them—any one of them—he would shout: "Tell Miss Scheremaier to come in." (No one can correct my spelling of the name for, to the best of my knowledge, this is the only time that it has been printed.) "Scheremaier" was also a name for inanimate objects, such as cans of paint and monkey wrenches. He would salute a small group of people as "Hello, Etcetera."

Professor March reported the following: One of the slower students was dozing, which Slichter noticed just as he was explaining about a body in a period of inertia. "Bill," he asked, "did you ever have a period of inertia?"

Typical also was his advice: "If you take the gum out of your mouth and use the other end of your head you will get the problem."

Professor Trumbower, a long-time family friend, recalls how the class sneaked two aces into Slichter's class cards. He began snapping them out on the desk calling the roll. When he got to the first ace he didn't bat an eye, he just called the name of the class dunce, his scapegoat for the semester.

Karl Paul Link, who is further described in Chapter 6, was a student of Slichter's in 1918–19. His description of surprise quizzes (as given in Mr. Allmendinger's notes) follows:

~ *"He came in, and we had the fear of God in us." Slichter came in . . . with a pack of blue books under his arm. He grabbed a handful and threw them up in the air, above the class. (This became a custom.) "Here," he shouted, "take one of these. I'm going to see how much you know." And proceeded to have them write their names and addresses and classifications in the blue books. That was the end of the examination. Other times he came in—unpredictably—with blue books, tossed them, said he was going to see how much they knew, and then say he'd changed his mind. Link says he was never sneaky on tests, but loved to play jokes on the class.*

As dean, Slichter vigorously sought grades from the most distinguished delinquent professors. Perhaps they could have benefitted by following his method of submitting grades. In a letter to President Fred, Rosetta Mackin (nee Powers), one of Slichter's former secretaries, relates: "When teaching Mechanics—went down to the Engr. Building, scratched one question on the board for the semester exam—told the students to write all they knew re same and bring their bluebooks up to the office when they finished. These students came up after slaving for 2 plus hours—I'm sure his grades were all turned in before the exam."

A teacher has relations with students which persist long after they have left his class. Some of these are official, such as writing recommendations; some are ceremonial, such as speaking at alumni gatherings; some are deepening friendships between persons who have found companionship in each other. All help to bind the alumni to the institution; but even more important, they enrich the lives of both the teacher and his former students.

In all these respects Slichter's outgoing nature made for more than usual activity. We will not describe but we will illustrate the above.

There is a recommendation, written in 1907, for a young man in the department, but I judge not actually one of Slichter's students; yet it shows the care and consideration given these matters.

~ *Mr. [X] has specialized in mathematics and physics and is a man of good scholastic ability, altho not brilliant. He will have no difficulty in getting his master's degree in mathematics at the University of Wisconsin upon the completion of his thesis. . . .*

I have not inspected Mr. [X's] work of instruction in the Madison High School; my judgment of his work is based on the many good things I have heard about his work and on the fact that I have had children in his classes in the high school from all of which I have heard most favorable accounts both as to his ability as an instructor and his skill as a disciplinarian.

When he wanted President Frank to read carefully a letter he had recently received, he wrote on October 12, 1932:

~ *I enclose herewith a document sent me by Mr. Glenn B. Warren, one of my former students. I place Mr. Warren as one of the six ablest students I have known at Wisconsin in my forty-eight years of service. He is now the principal engineer of the General Electric Company in charge of steam turbine development. . . .*

To me one of the most interesting features of his document is the fact that he, as an engineer, has been thinking seriously and deeply of our present economic difficulties.

It would not surprise me if Slichter showed equal enthusiasm for more than 2 × 3 persons. (Six is a nice number and it does not look quite as round as ten.) However, Glenn Warren belongs in very select company.

I have found practically no unfavorable letters which he wrote concerning any of his students. There are, of course, many possible explanations, and the combination of some of these probably existed: (1) I guess few students left his class without a pretty accurate idea of what he thought of them and, if the opinion was poor, would not initiate requests for recommendations; (2) Slichter was really a very kindly man and could, I believe, forget to write an uncomplimentary letter; (3) He had rose-tinted glasses when he looked at young men, especially when they were his students; and even if he could not turn pyrite into gold, he could believe he had.

It is only in light of some of his more enthusiastic letters, such as that mentioned in Chapter 6 concerning Max Mason's qualifications for the presidency of the University of Chicago, that one can understand how relatively restrained some of his other recommendations were.

Slichter was continually being asked to make short talks on various occasions, among them class reunions. Of course his old students flocked to hear him. George Haight, on February 28, 1934, sent his classmates the seventh notice of the class breakfast expressing his delight that Slichter would be with them and enclosing a copy of Slichter's letter of acceptance.

~   Dear George:

I am horrified to learn that it is proposed to bring that class of '99 back to Madison this coming June. I remember well how the Faculty struggled to get that class out in the world, stretching all University regulations to the limit, because we knew that you were the last representatives of the Victorian Age. We were determined to start the Modern Age next year with fresh material—it turned out to be plenty fresh—and start a new era. I suggest that you have your reunion in 1999—you all should be tamed by that time.

Nothing in my suggestion should, of course, apply to the women of the class. I well remember their universal charm, and how they were all marked excellent in mathematics. The reunion would be great if confined to them. I would be only too glad to serve as chaperon.

Sincerely,
Charles Slichter

From the place in the file where the following was found, I believe it is what Slichter said at the breakfast. At any rate, it is an orphan that must find a home.

~ *I am at a great disadvantage in speaking to you. Some think that the first signs of age are the miseries creeping up the backbone. But that is not the case. The first sign of the years is the loss of vocabulary, especially the loss of strong and emphatic and pungent words. Swear words are the first to leave us. It is a terrifying experience to find one self reaching for a cuss word and finding that it is not there. When we begin life, our first screams as infants are merely an instinctive groping for blasting words. This part of our education makes rapid progress and we soon acquire a rich vocabulary, only to see it vanish again later on. It is not the gout or wheel chair that is the symbol of age—it is the softening of the words. As our hair turns white, our language turns white also. We not only lose our molars and our incisors, we also lose our cuspids and bicuspids, so that we find ourselves unable either to cuss or to bicuss. Our friends soon note our loss of sulphurous utterance; the wife and the dog are especially conscious of it. They naturally ascribe it to the mellowing of age, but that is not the case— it merely means that the cortex of the brain has become emeritus.*

It is a serious predicament to have to face the world without a stout vocabulary. It brings on a gloom and closes easy avenues of escape. Man is not equipped to face life with only the language of the lily. All this means that there is left to me only the use of soft and kindly words. I can only praise you and acclaim your virtues and your marvelous accomplishments. But if I still possessed the vocabulary I once had, I could speak the truth and recite facts. Now I must flatter you and say nice things. You have given us a fine party and I am having a damn good time.

The above is an amplified echo of a conclusion he reached years before when, after listening to the linguist, Professor Edward T. Owen, he declared: "Owen made us all feel quite humble as we listened while one after another of our pet expressions were put on the operating table of his literary clinic. After due meditation, I have concluded that the only safe mediums of expression that gentlemen should be trusted with, are the symbols of the sign language and a full vocabulary of swear words."

The friendships with students are also described in other parts of this biography; but the mention of such names as Warren Weaver, Max Mason, and George Haight will indicate what is meant. Slichter could scare people when they first met him, but he possessed the alchemy that could change fear and respect into love and respect.

Textbooks

Although there is no question that one of Slichter's main purposes in writing textbooks was to make money, it would be unfair to describe this activity only under the heading of "Business," for one of these texts, *Elementary Mathematical Analysis,* was of such exceptional merit as to make a marked contribution to the teaching of first-year mathematics to engineering students, including those at Wisconsin. It also reflected the then-novel methods employed in this course.

In introducing his widely used *Introduction to Mathematical Analysis,* Frank L. Griffin (whom Slichter once tried to get to come to Wisconsin) wrote in 1920: "Under the traditional plan of studying trigonometry, college algebra, analytic geometry, and calculus separately, a student can form no conception of the character and possibilities of modern mathematics, nor of the relations of its several branches as parts of a unified whole, until he has taken several successive courses." In his text Slichter had already done an excellent job in *Elementary Mathematical Analysis* of unifying the first three of these subjects.

Few books give a better idea of what can be done with relatively elementary mathematics. Somehow the feel of applied mathematics comes through, especially perhaps in its emphasis on transformations and changes in scale. The treatment aids the development of one's mathematical intuition and imagination, a more difficult attainment than the acquisition of proof-making techniques. Slichter's engineering slant is brought out by an extremely didactic introduction. We find in the second edition:

~ *All mathematical work, such as the solutions of problems and exercises, and work in computation should be done in ink. The student should acquire the habit of working problems with pen and ink. He will find that this habit will materially aid him in repressing carelessness and indifference and in acquiring neatness and system. . . .*

Successful intellectual work depends very largely upon the power of concentration. Fortunately this power can be acquired and cultivated. The student should study away from interruption and then must not permit his work to become interrupted by himself or by others. By holding his attention upon his work and by keeping his mind from wandering to extraneous matters, the student will cultivate a fundamental habit that will tend to assure his success both in and out of college.

Professor March, many years after he had retired, spoke in high praise of this book which he so often had used and which prepared students to study calculus from the text he and Wolff had written, in a series edited by Slichter. I found a combination of *Elementary Mathematical Analysis* and Silvanus Thompson's famous, and at that time anonymous, *Calculus Made Easy* remarkably suited for a course aimed at preparing mathematically naive students, including faculty members, for their first work in mathematical statistics.

Professional Societies

An active member of a department naturally joins professional societies. Slichter did. When, in 1891, the New York Mathematical Society became national in scope and even before it changed its name to the American Mathematical Society, he became a member. He does not appear to have been very active in most of the numerous professional societies to which he belonged. However, he was a charter member of the Mathematical Association of America and was an associate editor of its journal, *The American Mathematical Monthly*, during 1915 and 1916. Moreover, from 1900 to 1902 he was president of the Wisconsin Academy of Sciences, Arts, and Letters. His presidential address, given December 26, 1902, entitled "Recent Criticism of American Scholarship," created widespread interest. He admitted the charge that the United States was far behind European countries in pure science, but pointed with pride to American accomplishments in engineering and applied science. He indicated that American technology was moving away from the empirical and was increasingly based in theory. This trend could be expected to increase American interest in pure science. He also declared that "the advancement of knowledge is as much a function of the university as is the propagation of learning." He ended with the plea: "Let all the productive intellectual forces of the state be united in this society as an instrument for the advancement of investigation and the spread of knowledge."

Chairmanship

As already mentioned, Van Velzer resigned early in 1906, and Slichter immediately became chairman. He was faced with a problem and an opportunity. From his actions we may assume that he had the assurance of the administration that a top-level mathematician could be added to the department.

Appointments. In 1906, except himself, there was no professor or associate professor in the department; there were two assistant professors: E. B. Skinner and L. W. Dowling.

Ernest B. Skinner was forty-three. He had come to the University in 1892 and had taken his Ph.D. in 1900 at the University of Chicago in group theory, a branch of algebra. He was an excellent teacher for all but the better students—exasperatingly slow from them. His patience and his ability to work were endless. He was moderately tolerant of the weaknesses of others but a strict puritan in his own behavior. He wrote a number of textbooks, one of which, *The Theory of Investment,* was a pioneer in its field. For decades he was in charge of the elementary mathematics for students in the College of Letters and Science. He never became a research mathematician. An admirable citizen, he played an important role in the creation and nursing of the State Teachers Retirement System; and for years he was the wise and firm president of the Madison School Board. He remained at the University until his death in 1935.

Linnaeus W. Dowling was thirty-nine years old and was on leave of absence studying at Turin University, Italy. He had taken his Ph.D. at Clark University in 1895 and had come to the University of Wisconsin in the same year. His field was geometry and he had just completed some work on conformal representation. He was also interested in actuarial mathematics. His mathematical production was not great; however, he was an interesting and lucid teacher and was well liked by both his students and his advisees. He died in 1928.

Of the junior staff, only Henry C. Wolff in applied mathematics and Florence E. Allen in geometry had been in the Department of Mathematics for a considerable time. Miss Allen had joined the staff in 1900 and was still active when Slichter retired.

The department had no specialist in analysis—at that time perhaps the most important branch of mathematics—and no member who had established himself as an investigator in pure mathematics, and only Slichter in applied mathematics.

On January 18, 1906, Slichter wrote to Max Mason, one of his former students, who had received his Ph.D. at Göttingen in 1903 and who was then in the Sheffield School at Yale:

~ *You have probably heard, from other sources, of the resignation of Professor C. A. Van Velzer from the chair of Mathematics at the University of Wisconsin. This gives us at Wisconsin the opportunity of a generation to do something for our much abused department of mathematics. We shall make a thorough canvass for the very best man available in this country, Great Britain or Canada.*

I shall be very glad to receive from you any suggestions that you *can make. We need a man who can take charge of the department and put it on the highest possible plane as a productive department, as well as an attractive one in the university. The salary that we can pay will probably not exceed $3,000 or $3,500 per annum. The university income is now ample for actual needs, and our library, as you know, is getting into shape, so that work can be done at the University of Wisconsin as well as at most places. I will not feel satisfied in hunting for a man for this place if we do not canvass the situation, both in Canada and Great Britain, as well as in the United States. I shall try to get down East and look over the situation there and also talk the matter over with yourself and your friends.*

It must be remembered that the number of Ph.D.'s granted in mathematics in the United States before 1906 was small. The University of Wisconsin had granted three. A few mathematicians had gone to Germany to study and their influence in America was to be great. Among these were William F. Osgood, who had received his Ph.D. at Erlangen in 1890, and Maxime Bocher, Ph.D., Göttingen 1901; both at Harvard and probably unmovable. The able American mathematicians of the right age who might be enticed to come to Wisconsin were few, but they were good.

Slichter consulted with a number of persons, particularly Osgood and Frank N. Cole—for years Secretary of the American Mathematical Society. In addition, President Van Hise received independent advice from Guy Stanton Ford, a graduate of Wisconsin, at that time a professor of history at Yale (later president of the University of Minnesota), who wrote a remarkably perceptive letter for one not in the field of mathematics.

Slichter made up his mind that the position should be offered to the English mathematician, Edmund T. Whittaker, for a four-year period although, if Whittaker would prefer, Slichter was willing to offer a permanent position.

Whittaker, thirty-two years of age in 1906, was to have an eminent career and was to be knighted in 1945. He had been Second Wrangler at Cambridge and was, in 1906, Fellow of Trinity College. From 1912 to 1946 he was a professor at the University of Edinburgh. His field was analysis on its applied side; he was interested in astronomy, interpolation theory and theoretical physics. He died in 1956. If he had developed in America the same interests he had in Britain, he would have added strength but not balance to the department.

I do not find any evidence as to what happened to this proposal. The offer may have been made and turned down; or Slichter may have found that Whittaker was unwilling to consider the proposition. In any case, nothing came of the matter.

In Slichter's mind the choice soon narrowed down to four persons: Henry S. White of Vassar, Gilbert A. Bliss of Princeton, Max Mason, and Edward B. Van Vleck of Wesleyan.

White was born in 1861 and had received his doctorate at Göttingen in 1891. He had gone from Northwestern to Vassar in 1905. His field was geometry. Later he was president of the American Mathematical Society, editor of its *Transactions*, and a member of the National Academy. He is remembered with affection by a host of Vassar graduates. Although he was an able person, he would not, in my judgment, have built up the department at Wisconsin as effectively as Van Vleck did.

Bliss was a younger man, then thirty years of age, and had received his doctor's degree from the University of Chicago in 1900. He was a preceptor at Princeton. Later he was to serve for many years at the University of Chicago. He, too, was to be president of the American Mathematical Society and a member of the National Academy. His field was analysis and his specialty, the calculus of variations. He was a fine mathematician with both administrative and athletic ability. He was a man of great charm, and was a friend of Max Mason. At the time, he was engaged to teach in the summer school at Wisconsin, and Slichter admitted to Mason that there were hopes that Bliss might sometime join the department. The nearest he came to this was to receive an honorary degree from the University of Wisconsin in 1935.

The general advice that Slichter got was to try to secure either Van Vleck or Mason. Osgood wrote: "It seems to me that you have to choose between a man who has already achieved a national and an international reputation and is still in his prime,—and one who is just entering on his career with the brightest of prospects." Ford also placed these two men at the top, rather favoring Mason.

Mason, a Wisconsin graduate, had recently gotten his Ph.D. at Göttingen. It was clear that Slichter would have been delighted to have him at Wisconsin. Mason was brilliant, was interested in applied mathematics, and there could hardly have been a more cordial relationship than that which existed between himself and Slichter (described in Chapter 6).

Edward Burr Van Vleck, born in 1863, was the son of John Monroe Van Vleck—an able mathematician and a leading faculty member at Wesleyan University. The younger Van Vleck received his doctorate from Göttingen in 1893, and then became instructor at Wisconsin for two years. In 1895, he returned to his alma mater, Wesleyan, where he became a professor in 1898 and a colleague of his father, who retired in 1904. He was to be a colleague of his son, John Hasbrouck Van Vleck, at the University of Wisconsin. Like White and Bliss, E. B. Van Vleck was to become president of the American Mathematical Society and a member of the National Academy. His field was analysis.

Slichter finally laid the choice before either Birge or Van Hise, and the final decision to offer the position to Van Vleck was made by the president. It was a wise decision for, although Mason would also have been an excellent choice, Van Vleck was established and brought recognized strength to the department, as well as giving it balance among the fields of mathematics. Moreover, the president may have recognized that Slichter would never relinquish the idea of bringing Mason back and may have foreseen that it would be possible to do so, for in 1908 Mason returned to the department as an associate professor. In 1910, in accordance with his own wishes, Mason transferred to physics as Professor of Mathematical Physics with the approval of Van Vleck who wrote Slichter, then abroad, "The taste of so brilliant a man as Mason should be complied with as far as possible." The appointment of Van Vleck was the most important in the history of the department since Slichter had been invited to join it by Van Velzer, and of course was made with much more realization of its significance. Yet one could hardly have known in 1906 that on his death the faculty would say of Van Vleck: "The eminence of any institution is based on a small number of such men. Professor Van Vleck brought lustre to the University of Wisconsin."

The appointment of Herman W. March was made in the same year. He came to replace Peotter, whose engineering work for Slichter is mentioned in Chapter 4. Born in 1878, and graduated from Michigan in 1904, March had taken his master's degree there in 1905, and in 1906 was an instructor in physics at Princeton. He had been an assistant in astronomy at Michigan. He left Wisconsin in 1909 to study at Munich, where he received his Ph.D. in 1911. He then returned to Wisconsin and remained until his retirement in 1949. March succeeded Slichter in taking charge of the mathematics instruction for engineers and was consistently productive in the field of applied mathematics, not only before he retired but even after, when he served as consultant for the United States Forest Products Laboratory. His gentle tact was often to serve the department in good stead. He lived until the autumn of 1969.

Slichter had the unbeatable combination of sound judgment and good luck in the appointments to the department made in his first half-year as chairman. It is difficult to assign credit for the appointments that were made after Van Vleck came. It can be seen from the already quoted letter to Mason that Slichter expected that the new professor not only would be a research man and teacher of advanced students, but would also take a leadership role in shaping the destiny of the department.

The letters of both Van Hise and Slichter offering the position to Van Vleck are lost or destroyed. We have Van Vleck's reply to Slichter, written in Venice on April 10, 1906. From it, much of what Slichter must have written—even some of its flavor—can be deduced. The major portion follows:

~   *I suppose that the very best answer to your cordial letters is an acceptance of the Wisconsin offer. This acceptance I sent yesterday by cablegram to President Van Hise, and now I have the pleasure of replying further to your letters. Possibly I should have written earlier, but with so momentous a question before me for decision, my thoughts have been engrossed with its consideration rather than with letter writing.*

It is, of course, a great personal gratification that you who have known me at Wisconsin have been so largely instrumental in drawing me back, and it is my earnest hope that you may not be disappointed in the future.

Your words regarding the increase in the mathematical equipment of the University and in the mathematical appropriation were especially welcome. Indeed, it is the advanced character of the work possible at Madison which is especially attractive to me. I have already begun to think and to plan for next year's work! In particular, I am wondering whether the University has a set of the Fortschritte der Mathematik, which is to me almost indispensable. I do not remember to have seen it two years ago.

You are entirely right in saying that you could think of but few places in the country where I would be so happily domiciled as at Madison. Indeed, there are only a very, very few places which could have attracted me away from Middletown, where we have lived so happily. My recollections of the beauty of Madison were only enhanced by my visit two years ago.

From 1906 on, both Slichter and Van Vleck were active in searching for and selecting staff members. For the two years from 1908 to 1910, when Mason was in the department and after Skinner became associate professor in 1910 and Dowling in 1914, the responsibility would have been shared (I fear to a rather small degree) with these men.

At times the department operated in two almost autonomous divisions, of pure and of applied mathematics. In 1905 the stationery used by the department carried the names of the staff members in two columns on the letterhead; and although these lists had no special designation, it is clear that the one led by Van Velzer was the staff primarily teaching Letters and Science students and the one led by Slichter, Engineers. Slichter's official title was Professor of Applied Mathematics. The 1906–7 departmental report was in two parts, the second written by Van Vleck. Negotiations with young staff members were sometimes carried on by Slichter and sometimes by Van Vleck. When I first came to Wisconsin all my contacts were with Professor Van Vleck.

The following persons who reached the rank of assistant professor or above at the University of Wisconsin, and whose first appointments were between 1906 and 1920, when Slichter was chairman (except for the year 1909–10 when he was on leave of absence and Van Vleck was chairman), are listed with the years when their names first appeared in the catalogs of the University of Wisconsin as members of the staff of the Mathematics Department: Edward B. Van Vleck, 1906; Herman W. March, 1906; Max Mason, 1908; Horace T. Burgess, 1909; Arnold Dresden, 1909; Walter Hart, 1910; Eugene Taylor, 1910; Rodney W. Babcock, 1916; Warren Weaver, 1918; Ernest P. Lane, 1919; Mark H. Ingraham, 1919.

It was Lane who credited his experience at Wisconsin (although he had already taught for six years) as teaching him much that he used for years thereafter, and he stated that the standards of freshman engineering mathematics were the highest of any in which he had participated.

Among others appointed while Slichter was chairman were: Thomas M. Simpson (1908–18), later chairman of the department and graduate dean of the University of Florida; Elmer E. Moots (1909–11), who worked in mathematics at Cornell College and sent a sequence of first-class students to the University of Wisconsin for graduate work; and Ottis H. Rechard (1919–23), who built the department at the University of Wyoming and became dean of the college at that institution.

This is a strong record and, at least in the cases of Van Vleck, March, Mason, Weaver, and almost certainly some others, the initiative was Slichter's.

Space Problems—"*For men may come and men may go, but I go on forever.*" In the fall of 1906, President Van Hise addressed a circular letter to the various departments, asking them how much space they were occupying at that time and also one and two decades earlier, as well as affording them an opportunity to present their needs. Considering that, in 1920, when the Mathematics Department took over North Hall, and again, in 1963, when Van Vleck Hall was completed, solutions of the departmental space needs for the then-immediate future were found, Slichter's analysis is of interest. Excerpts follow:

~ 1. *There should be at least one office for each two professors or assistant professors in the department. There should also be at least one office for each three instructors.*

2. *The recitation rooms for all of the elementary work in mathematics should contain not less than 800 square feet in order to provide adequate space for blackboards. . . .*

3. *The rooms used for mathematical instruction should be more compactly arranged than at present so as to secure the solidarity of the department and to render communication between instructors as convenient as possible.*

Under the head of present needs it should be emphasized that the present accommodations of the department of mathematics in the University Library are not what they should be, even under existing conditions. At the present time we occupy one poorly lighted room in the Library as a seminary room, and this room contains all the mathematical books and current periodicals. At least one additional room should be at hand which could be used as a reading and research room for the students and members of the faculty. Adequate provision for the needs of the department would include not only a reading room for students but also a small room in conjunction with the seminary room which could be used for research work by members of the instructional staff.

He then continued the discussion under the heading: "Permanent Housing and Natural Affiliations." He stated that the closest connections of mathematics are with physics, but that the types of buildings needed by physics and mathematics are so different that, although near, they should be separate. He pointed out the similarity of the rooms needed for language instruction and mathematics, and the large number of students who take both. He concluded this section with:

~ *The above paragraphs indicate that the best provision for the department of mathematics would undoubtedly be made in a building especially constructed for their own needs, conveniently adjoining the buildings given to language and the humanities, and that next to this ideal provision is the provision for the department of mathematics in the same building or buildings with the literary departments of the College of Letters and Science. It should be noted, however, that if the*

College of Engineering should be removed to a position more remote from University Hall than it occupies at present, it would undoubtedly be best to provide for the mathematical instruction of engineering students in a building convenient to the principal engineering building.

His forecasts follow:

~ In projecting into the future the requirements of the department of mathematics it must be remembered that the principal growth must be in the elementary courses. The number of sections in elementary mathematics must necessarily increase with the increase in the number of students, but the number of sections in advanced mathematics would not greatly increase although the number of students taking such courses will gradually grow. At the present time there are ten sections in mathematics of the junior year and above in which are enrolled about 70 students. Twenty-five or fifty years from now the number of sections in advanced mathematics may not increase at all; possibly there may be twelve or fifteen sections of this grade and 200 or 300 students in the classes. Such change will not greatly affect the new needs for space in the future.

And again,

~ . . . In 1956 the estimate indicates the necessity for twenty-one instructors, 58 sections in elementary mathematics, and 12,000 square feet of floor space.

The plea for space got unusually swift action, for in less than fourteen years, as one of a series of moves of departments made possible by the completion of Sterling Hall, mathematics took over all but three rooms of North Hall. This gave them fourteen classrooms and twelve offices. North Hall contained about 12,000 square feet of assignable floor space, had already been used by languages, and was located between the Engineering Building and University Hall (Bascom Hall), where the humanities were. Quite according to the doctor's orders! The location in North Hall also conformed to the erstwhile fashion in the United States of housing mathematics in the oldest building on the campus.

Like many another educator dealing with enrollments and space, Slichter proved to be a poor prophet, although he did indicate that he considered his estimates conservative. In the fall of 1956 the enrollment in elementary courses in mathematics was 3,117 taught in 118 sections; in advanced courses in the junior-senior-graduate level, it was 672; and in purely graduate courses, 257. The numbers of distinct courses in these last two groups were 12 and 11, respectively, taught in

30 sections. The staff of the department consisted of 20 persons above the rank of assistant and a host of assistants who would have been called "instructors" in 1906. The department was busy planning a building—which turned out to have 107 offices, 32 classrooms, and 57,580 square feet of space, not including halls, stairways, and rest-rooms.

What you have must be guarded and kept in order. In the fall of 1906, the political scientists broke the Tenth Commandment, and the mathematics seminar room was in jeopardy. Professor Paul S. Reinsch wrote to the president to the effect that this room had been assigned to political science and loaned to the mathematicians and that the time to call the loan had come. He referred to the room as the "Political Science Seminar Room." On October 2 Slichter wrote to the president:

~ *I have your note inclosing letter of Professor Reinsch, dated Sept. 6th, in which he requests the use of the present mathematical seminary room in the Library. He states that the two rooms at present assigned to the Department of Economics and Political Science are not adequate for their purposes.*

I have a very high opinion of the Department of Political Science and Economics and appreciate the fact that they are desirous of securing all the available space possible for their work. However, it must be remembered that the present seminary room in the Library building assigned to the Department of Mathematics is the only home this department has. If we had two rooms it might be practicable to consider surrendering one of them temporarily. The seminary room contains all of the bound periodicals of the mathematical department, all mathematical books and treatises and all the current numbers of mathematical periodicals. The room is open for the use of all members of the mathematical instructional force, all fellows and scholars or graduate students, and advanced students of junior and senior grade. In addition to this, during the present year seminary courses are conducted in this room by Dowling, Van Vleck and myself. The number of advanced students in the mathematical courses is much larger this year than it has been in the past and we feel that we are just entering upon a period of increased usefulness and scientific activity. I believe if Professor Reinsch realized the impossibility of reducing us from one room to no room, he would not make the request that he has. It would be very much easier for the Department to reduce its rooms from two to one than to reduce from one to absolutely none.

I shall be very glad to add any further particulars in case you desire them.

This was effective, for mathematics kept the room for many years until a Physics-Mathematics Library was established in Sterling Hall with a librarian in charge. It is now in Van Vleck Hall.

This episode evidently put the room on his mind, for on October 9th he wrote the president (why not the librarian?) suggesting that the janitor get to the building fifteen minutes before it officially opened so that the seminar room could be available for immediate use at 8 A.M.

Not only space but the condition of space stirred Slichter. For instance, the superintendent of buildings received a letter, written March 22, 1907, stating: "Two of our instructors have been moved from Room 60, N.H., in which the chalk trays were cleaned from dust twice a day, into rooms in University Hall where the chalk dust is allowed to accumulate and the air is so filled with dust that the instructors have serious trouble with their throats and have been obliged to lose several days work." On November 12, 1908, he wrote to the chairman of the University Committee on Hygiene, mentioning the above complaint and reporting a decided improvement in that respect. But he said:

~ We are suffering this year a great deal in University Hall on account of the lack of ventilation. . . .

In general I find, however, that it is very difficult to get the janitors to look upon the forced ventilation system as anything else than an auxilliary to the heating plant. They always get the impression that the only need of working the fan is to furnish heat and that moderate weather does not require the use of the fan because the direct radiation will furnish sufficient heat. If our janitor continues to adhere to this ancient theory, I hope that your committee will be able to show him the necessity of running the fan in mild weather and giving ventilation to all of the rooms.

The real explosion came in 1915. On October 29 the following letter, clearly written by Slichter but also signed by the other professorial members of the department, was sent to Van Hise:

~ At the opening of the college year the members of the mathematical staff were surprised to find that a large part of rooms 203 and 252 in University Hall had been cut off and converted into toilets for students. When we became aware of the situation, the work had already progressed so far that it was impossible to have it stopped. The department wishes nevertheless to enter a formal protest against the action that has so seriously interfered with the usefulness of its rooms.

As far as we know, there has been no consultation of any sort with any members of the mathematical department by those persons who may be responsible for the execution of the change. As our department is most vitally affected, we think that the change should not have been made without the courtesy of prior information and the opportunity of a conference. As long as can be remembered, these rooms have been assigned to the mathematical department, and they have been its best rooms because of peculiar fitness for mathematical work. It seems obvious that the method of making changes pursued in this instance would, if prevalent, be unfortunate not only for particular departments, but for the University.

From the standpoint of our department the change seems regretable and injurious. Until now, the rooms have been especially valued for meetings of large mathematical groups and classes, for which we have no other good rooms. They now present a cell-like appearance and have lost their cheer. Few departments are so vitally interested in the amount, convenience and lighting of its blackboard space. In these particulars the rooms have been badly damaged. The noise from the flushing of the adjoining toilets is noticeable and objectionable. The ventilation has also become difficult.

Into a discussion of the suitability of the situation selected for student toilets we do not propose to enter other than to suggest the following considerations: The toilets are in the middle of the building just where the congestion between classes is greatest. The only door from either toilet opens into the main corridor so that every person who enters will publish that fact to the other students, both male and female. The indelicacy of proximity of the toilets for the two sexes is already a matter of some comment. If the courtesy of conference with several departments had been extended, it seems probable that some other more suitable provision could have been devised.

Van Hise's reply was a model for administrative brush-offs:

~ A copy of the letter of the mathematical department in regard to additional toilet facilities installed in University Hall was sent to the business manager and a copy to the chairman of the committee on rooms and time tables. I enclose herewith copies of their replies.

So far as I can see it is useless for me to take time personally to go into the merits of the question, since the work has been done. However, in the future I shall try to see that the department concerned is consulted in case of changes which may seriously affect it.

Protection of Staff. Slichter was ever ready to level a potent lance in defense of his younger colleagues. In 1908 Slichter had recommended a hundred-dollar increase for each of several staff members, including

Herman March. I judge that he had gone somewhat over the amount that was available for mathematics, for without consulting him the administration deleted March's increase. It was inevitable that Slichter would declare—and in this case accurately—that March should have been the first to get a raise rather than the first to be omitted. His reaction was prompt.

~ In making recommendations for compensation of instructors in the Department of Mathematics for the coming year, I included additional compensation of $100 per annum for four of our instructors who handle the work in mathematics for students in the Engineering school. . . .

I have just been informed that the additional compensation of Mr. March has not been allowed, but that the additional compensation of the other instructors has been allowed as recommended.

I very greatly regret this change, and Mr. March feels the discrimination very deeply. This is aggravated in this case by the fact that Mr. March is one of the very best instructors we have, and one who deserves the increase in compensation more than any of those who received it: if the compensation of any of the instructors was to have been left unchanged it should have been that of Mr. [X], whose work is not nearly so successful as that of Mr. March. . . .

I write this note as it is only just to Mr. March that I should explain the matter very fully, and if there is any way still available in which his salary can be increased, I hope it will be done. If it were practicable to take $100 from the salary of Mr. [X] and add it to that of Mr. March, I would recommend doing so, were it not for the fact that Mr. [X] has already been advised of his advance.

All of the young men recommended for the increase in salary thoroughly deserved the encouragement and reward for faithful service, and it only happens that the one who deserved it most was the one who, by accident, seems to have suffered in the revision of the budget.

March got the raise.

On another occasion, after Slichter had become dean of the Graduate School, George C. Sellery, Dean of the College of Letters and Science, maintained that it had been agreed that the appointment of a young staff member in mathematics had been renewed for one year only; and, although he continued it beyond the year, he refused salary increases and finally insisted that the appointment be terminated. However, a recommendation for a salary increase for the young man went to the Regents, not through channels, and this further raised Sellery's ire. President Frank was called upon to settle the matter. The salary increase was approved, but we have no record as to the agreement on terminating the appointment. It is true, however, that the member concerned soon went to head the department at another institution, and not long thereafter transferred to a public university, later to become dean of one of its colleges. A young man would think twice before challenging Slichter's authority as chairman; on the other hand, he could be sure that he had the protection of an effective champion.

Applied Mathematics at the University of Wisconsin.   It is due to Slichter more than to anyone else that the University of Wisconsin has had great strength in applied mathematics. As already detailed, he taught many of the more important branches in this field. He had distinguished students who also taught here—Max Mason, Warren Weaver, Thornton Fry, for example. He brought applied mathematicians, including Herman March, to the department. The tradition was continued until, partly because of this emphasis, the United States Army Mathematics Research Center was located at the University.

A document, prepared by Professor Rudolph E. Langer (later to become the first director of the Center), supporting the proposal that the Center be located at Wisconsin, describes the development of the department. He starts with the work of Van Vleck and Slichter and says: "Contemporary with Van Vleck was Charles S. Slichter, a mathematician of a different type. He was to round out Wisconsin's early reputation. Slichter was an applied mathematician. He founded a tradition of respect for this facet of mathematics here."

Relations Within the Department.   Internal tensions within departments are among the most common academic phenomena. The Mathematics Department during Slichter's chairmanship was not exempt. Both Van Vleck and Slichter worked conscientiously and successfully to develop a strong department. If Slichter was away, Van Vleck regularly wrote cordial letters to him concerning all major problems. Until the department occupied North Hall, they shared an office—254 University Hall. They had in common a love of travel, and this showed in their correspondence. However, they were not thoroughly congenial. A little of the stereotype caricature of the East versus the Midwest was shown in the contrast between the two. Each differed from the other in nature and in professional ideals. Van Vleck was reserved and quiet. His boyhood in an intellectual family, with a father who was a leader at Wesleyan University (a small Connecticut college—then, as now, excellent in its standards), intensified these characteristics. His training and interests were scholarly, and he delighted in the elegance of rigorous proof and careful exposition. Slichter's sudden decisions, loud voice, and at times preposterous statements could not help but jar Van Vleck's nerves. Van Vleck's disdain for some of the more vigorous, but to him vulgar, aspects of American life, was foreign to Slichter's robust nature. Slichter's emphasis on the use rather than on the development of mathematics would lead to differences with Van Vleck on policy. There was, for a short period, a tendency for the older members of the department to take sides on personal as well as other considerations. I believe the differences between others were sometimes sharper than those between the principals. March, at times, acted as conciliator.

Yet, after the above is acknowledged, the magnificent qualities of these two men were shown in their relations to each other. At no time did younger men feel under pressure to dislike one because they liked the other. Thus Warren Weaver, whose undergraduate teacher and hero Slichter was, developed admiration and affection for Van Vleck; and I, whose family background was rooted in Wesleyan and friendship with three generations of Van Vlecks and whose every contact with Van Vleck increased my friendship and admiration for him as a scholar, a teacher, and a man, came to care deeply for Slichter and to admire him also as a scholar, a teacher, and a man. The children of the two became close friends.

In those happy days when Van Vleck was chairman and Slichter was dean, tensions subsided. Important appointments were made on Van Vleck's recommendation and with the strong backing of the Graduate School. Finally, in the spring of 1928, Van Vleck called a departmental meeting; announced that he intended to retire in 1929; and characteristically stated that he believed it was not wise for a man to serve as chairman in his last year, making decisions with which others, but not he, would have to live. Slichter attended the meeting and at the end stated that he had taken great satisfaction in the development of the Mathematics Department, that it had become truly distinguished, and that the credit should go entirely to the leadership of Professor Van Vleck. This was the act of a warm and generous heart. It was also an act of academic statesmanship. I believe it moved Van Vleck deeply. It is my most vivid and happiest remembrance of the two men together.

In 1938 Slichter sent Van Vleck a copy of his volume of essays, *Science in a Tavern*. In reply Van Vleck said, "It is needless to say that I am enjoying and shall enjoy the book extremely, every word from end to end. I know your fine English and sprightly interesting style. The contents of course are of just the kind I like. You make one ashamed of not being polymath, like Sir Christopher Wren and some others I had not put into that category." This, too, came from a warm and generous heart.

Chapter 4 ❧ *Scientist*
and Engineer

BEFORE I describe Slichter's scientific work, I will make a few remarks on what various types of mathematicians do. Pure mathematicians produce new mathematical theorems, usually with proofs—although some conjectures, unproved at the time of publication and a few still in this state, have played an important role in the development of mathematics. The relationship of these theorems to each other is a major source of mathematical beauty. The content of these theorems is at times suggested by the desire to generalize already known mathematics or to solve mathematical problems that are presented by each new advance of the subject. At other times the mathematics is suggested by problems in other fields, such as physics, astronomy, or gambling—to mention three important sources of inspiration. Often the results are of greater mathematical interest than of importance to the applied field. Other mathematicians are more insistent that their results be applicable to the subject from which they sprang. A third important group of scientists, while thoroughly sophisticated in their mathematical knowledge, care most about the results of applying mathematics—frequently already known mathematics. Their originality (often no less than that of pure mathematicians) lies in their perception of the relation of phenomena to mathematical models. Slichter was one of these. His work was important, but its importance was to geology rather than to mathematics.

It is interesting to note that Slichter and some of his closest associates paid little attention to the barriers between disciplines. Slichter was in the department of mathematics, but his own technical contributions were in geophysics and hydraulic engineering; moreover, his interest in the history of science was broad even if largely confined to the physical sciences. Max Mason took his degree in mathematics, and his first appointment was in that department although he later held a professorship in physics, the field in which he did his most important scientific work. Warren Weaver took his degree under Mason in physics but was chairman of the Department of Mathematics at Wisconsin when he resigned to join the Rockefeller Foundation where, after years heading the division of natural sciences, he became a vice-president and gave a major portion of his attention to biology. This splendid disregard of conventional bailiwicks was a hallmark of Slichter and the men he influenced.

Scientist

Flow of Underground Water. Slichter's chief scientific work was connected with the flow of underground water. There were two phases of this work, overlapping in time and somewhat in subject matter: (1) the theoretical treatment published in monographs—mostly as reports of the United States Geological Survey; and (2) work as a consulting engineer on a great variety of projects concerning, for instance, irrigation, city water supplies, and legal disputes as to the effect of the use of water from one source upon the supply from another.

It would appear that Slichter became interested in this subject in about 1894 through collaboration with Professor Franklin H. King, a physicist in the College of Agriculture, who had been at the University of Wisconsin since 1888. He was sixteen years Slichter's senior, had worked on the Wisconsin Geological Survey from 1873 to 1876, and for the decade preceding his coming to the University had been a Professor of Natural Science at River Falls State Normal School (later Wisconsin State University—River Falls). He left Wisconsin in 1901 to become Chief of the Division of Soils Management in the United States Bureau of Soils.

In 1899, in the *19th Annual Report of the U.S. Geological Survey*, King published a major paper on "Principles and Conditions of the Movements of Ground Water." In the "Introduction" he described his work with Slichter:

~ *During our earlier investigations regarding the flow of water through soils, it appeared that if the laws of capillary flow apply to the movements of water and of air through soil, it ought to be possible to arrive at the sizes of soil grains from a knowledge of the flow of water through the samples under known conditions. Such great difficulties, however, were encountered in duplicating results with water that air was substituted as the medium whose flow was to be measured. The handling of the air proved so much simpler and expeditious and results could be duplicated so closely that in 1894 the plan was laid before Professor Slichter for his judgment as to the possibility of placing the method on a quantitative basis. This seemed to him possible, and he kindly consented to undertake a preliminary investigation, which resulted in the formula for computing the effective sizes of soil grains, presented in the first portion of his paper in this volume. When it was found that computed results agreed with observations more closely than had been hoped at first, a return was made to water as a means of checking the accuracy of the method and the formula. It was found that the flow of water used in the formula gave results quite comparable with those computed from air.*

At this stage Mr. Newell [F. H. Newell, for the United States Geological Survey] proposed, in 1896, to assist financially in an investigation of the movement of ground water, and the writer consented to undertake the work, with permission to secure Professor Slichter's services in the development of certain theoretical phases of the subject.

We have no record of whether King sought Slichter's help because of Slichter's general reputation; whether it was on the suggestion of Professor Stephen M. Babcock whom Slichter had aided on the "milk test"; or, perhaps even more likely, President T. C. Chamberlin (who knew Slichter well) had recommended that King consult him. King himself had been a student of Chamberlin at Whitewater Normal School (later Wisconsin State University—Whitewater) and had developed the work in soil physics at Wisconsin under Chamberlin's sponsorship.

Slichter's most important scientific paper was published in 1899 in the *Annual Report of the United States Geological Survey* (11, part 2, pp. 295–384) entitled "Theoretical Investigation of the Motion of Ground Waters." This is the paper referred to by King in the foregoing quotation.

Since this biography is not a technical one, the mathematical details of the work will not be given; however, a description of its nature and importance is appropriate. In this paper he made daring assumptions (most applications of mathematics are based on daring assumptions—Euclid's were heroic) to form a model tractable enough to be dealt with by mathematics but yielding results that can be translated into predictions and descriptions of the behavior of nature. The first of these was that water passing through gravel or porous rock would behave as if passing between a set of solid spheres all of the same radius, packed in some stable arrangement, the most interesting being the case of the most compact packing possible. The most open of these arrangements would yield the porosity of 48% and the most compact, 26%—limits between which the actual porosities of soils in general lie. With this background an elaborate discussion of the paths between the spheres is given, based on the assumption that valid results can be obtained from homogeneous packing throughout gravel contained in cylinders through which water passes. Slichter then proceeds to an approximation of water flowing through a tube whose cross sections are equilateral triangles. The results seem to have justified the assumptions, but it is hard to determine whether to admire more the accuracy of his intuition or the extent of his faith.

A second assumption, of less importance but in application very convenient, came from Professor King, namely, that the porosity of sand could be measured by forcing through it air rather than water, and hence the proper diameter for the spheres in the equivalent model could be determined.

To me, by far the most interesting portion of this paper is the short Chapter II entitled "General Laws of the Flow of Ground Waters." Starting with "Darcy's law" for the velocity of the flow of a liquid in a given direction through a column of soil (a law which had been in use for forty years), Slichter generalized it to three dimensions and showed that for incompressible fluids the pressure functions (p) satisfy the Laplace equation:

$$\frac{\partial^2 p}{\partial x^2} + \frac{\partial^2 p}{\partial y^2} + \frac{\partial^2 p}{\partial z^2} = 0,$$

an equation which Slichter called "the familiar equation occurring in nearly all branches of applied mathematics."

Slichter observed: "It seems remarkable that the fact that the solution of any problem in the motion of ground waters depends upon the solution of the differential equation [above] has not been pointed out before. The existence of this function is made the basis of nearly all the work in the following pages." And again: "We have, therefore, shown that a problem in the steady motion of ground waters is mathematically analogous to a problem in the steady flow of heat or electricity, or to a problem in the steady motion of a perfect fluid. The unknown function p must be determined from the partial differential equation, subject to the boundary conditions present in each particular problem undertaken. The function p is necessarily finite, continuous, and single valued at all points of the hypothetical liquid, and the general methods usual in potential theory apply."

From this point forward Slichter exploits, through the skillful use of mathematical methods developed during the nineteenth century, the consequences of the assumptions already made. In some cases the results were startling to many geologists. It had often been argued that the withdrawal of water would not interfere with the flow of wells in the neighborhood unless it was sufficient to lower substantially the water table. This is not in accord with either potential theory or experience.

Slichter says: "A remarkable fact brought out by this diagram is the distinction that must be made between the level at which moving ground water stands and the actual pressure to which the ground water is subjected at a given point. It is often assumed that the level of the surface of the water in a well indicates the position of the water table and that a change in the height of water in a well indicates the same change in the level of the water table."

And again:

~ We conclude, then, that the wells in a given district might be seriously interfered with by the construction of a drainage flume near them, and yet the water table in the neighborhood of the wells might show but little change of level. Nevertheless it is sometimes maintained that a drainage flume can not be the cause of the lowering of the water in the wells of a district unless the amount of water entering the flume in a given time after it is first used is enough to lower the entire water table of the district by an amount equal to the measured fall of the wells during that time. Thus, if the wells of a district containing one square mile are found to have fallen one foot after a new drainage flume has been used for a month, and if the water-bearing soil contains 20 per cent of water, then the argument is that the drainage flume has not caused the fall of water in the wells unless the flume has actually taken off $\frac{1}{5}$ of $5{,}280^2$ cubic feet of water in the month considered. A general lowering of even a few feet in the water table of a district represents an enormous amount of water, and a flume is seldom likely to be held responsible for an interference with wells if tested by the principle just mentioned. The lowering of the wells, although a fact, would have to be explained by the hypothesis of a "dry season" or by some other convenient hypothesis. The facts . . . show, however, the fallacy of the argument. The flume may not cause a lowering of the water table for any great distance from its walls, but the lowering of the pressure, as indicated by the lowered position of the water surface in the wells, may extend to great distances. This is a fact that nonmathematical reasoning shows must exist. The water supply for the *flume comes from the surrounding soil; hence the water in the soil is in motion. The motion of the ground water against the resistance of the capillary spaces of the soil requires the existence of an adequate moving agent. This moving agent must be a pressure gradient extending away from the walls of the flume. This pressure gradient must come into existence as soon as the water begins to move and before there is time for an appreciable lowering of the water table to take place. The drop in the water level in the wells shows the change in pressure of the water in the surrounding soil and does not necessarily indicate a change in the position of the water table.*

At the end of this paper there is a discussion of the mutual interference of (a) two wells, (b) three wells, and (c) a large number of wells in a row.

This masterly paper made a stir among geologists, hydraulic engineers, and irrigation experts. It led to Slichter's becoming a leading consultant on many matters, as will be described. Since its originality was in applying known mathematics to unsolved engineering and geophysical problems, it was scarcely noted by the mathematical fraternity. This fact is not derogatory either to their intelligence or to Slichter's contribution. It simply means that the chief significance of Slichter's work was to others rather than to mathematicians. As witness to its importance, it is interesting to note that in a later paper on capillary flow Edward E. Miller, of the Department of Physics of the University of Wisconsin and the physics consultant of its Agricultural Experiment Station, hence King's present successor, wrote in 1955: "By adding a conservation-of-matter condition to Darcy's law written in differential form, Charles Slichter in 1899 obtained an equation identical in form with the heat flow equation. This has since formed the basis for the successful development of saturated flow technology."

In 1902 Slichter wrote a remarkable sequel to his 1899 paper. This was published by the United States Geological Survey as "Water Supply and Irrigation Paper No. 67," under the title, "The Motions of Underground Waters." It would be difficult to find a better example of semipopular exposition of a technical subject. In his letter of transmittal, F. H. Newell, the head of the Division of Hydrography, wrote the Director of the Survey:

~ . . . *Professor Slichter has been giving considerable time and attention to experimental and theoretical considerations of the movements of underground waters, the preliminary results of which were published in Part II of the Nineteenth Annual Report. The present paper treats of the simpler and more general topics connected with the* movements of water underground, being intended to answer the more elementary questions which arise in a consideration of the subject. Examples are given of the various areas in which water occurs underground, the origin and extent of the waters are discussed, and methods of bringing them to the surface and making them available are touched upon.

The chapter headings are: i. Origin and Extent of Underground Waters; ii. Surface Zone of Flow of Ground Waters; iii. Deep Zones of Flow; iv. Recovery of Underground Water from the Surface Flows; v. Artesian and Deep Wells.

Slichter reviewed the state of knowledge at the time, working his own contributions both theoretical and instrumental into their logical places without either undue emphasis or false modesty. At least once his enthusiasm dominated his style when he wrote of "a very unique and not uncommon phenomenon."

I have seldom learned as much in twenty-four hours as during the day, July 15–16, 1969, when I read this paper. It is a wondrous tale—the surface river and the plane, the seeping down of rain, and the capillary upflow in periods of drought, but chiefly the great bodies of water accompanying the flow of rivers or underlying at moderate depths semiarid regions, or water in porous rocks below impervious shale or other stone shields, where the rock above and the water below may hold gas and oil fields captive for eons, and where water, traveling perhaps a mile a year, supplies our artesian wells. When needed, theory is introduced or referred to; little new theory is reported, but this is the first general paper in which I have found a description of the improved methods developed by Slichter, starting in 1901, for measuring the direction and velocity of the flow of water. The older methods introduced some chemical (for instance, salt) in an upper well and by a sequence of chemical analyses determined when the solution had reached a lower well. Slichter introduced an electrolyte into the upper well and measured the electric current, both between and in wells, instead of making a chemical analysis. This was not only a quicker process but one well adapted to recording instruments.

Anyone from a twelve-year-old to a Nobel Prize winner would be fascinated by this paper, though the former might wish to omit the few formulae that occur, and the latter the description of flushing the wells in Savannah. Those of us between can enjoy both.

In an address given in 1969, I said: "A disproportionate share of our scientific energy is going into seeking new knowledge as compared to organizing the known. I believe that one basic reason for this is that discovery is easier than organization, yet, strangely, it receives more honors." It is pleasant to find one's advice taken sixty-seven years before it was given.

Slichter sometimes used reports on a particular project as a vehicle for describing a general method. Thus in 1904 he published in the journal, *Western Society of Engineers,* a paper entitled "Measurement of Underflow Streams in Southern California," in which the most interesting and longest section is a "Description of Underflow Meter, Used in Measuring the Velocity and Direction of Movement of Underground Waters," a detailed description for engineers of the method given in the 1902 paper, No. 67, already mentioned. His enjoyment of the actual problems of work in the field is illustrated by detailed advice as to the types of pipes (one and one-half inch better than one and one-fourth inch because the former is lap-welded and the latter butt-welded), the drive points, the methods of pulling the casings, the electrolyte (sal amoniac), and a multitude of other items dear to the heart of a man who was not only a scientist but a born gadgeteer.

Slichter's satisfaction in the development of the method for measuring the flow of underground water was shown on later occasions. As illustrations: "My method of measuring the rate of underflow gives autographic records made by recording instruments. These original records can be placed in the hands of capitalists or investors so that they know what they are getting. There is no need of 'guessing' about the amount of the underflow as we measure the same about as accurately as a surface stream can be gauged, and then turn the automatic records over to the interested parties upon which they can base their scheme of development." And in 1912 he wrote: "I have devised an instrument for measuring the rate of movement of underground waters, which last had made me a pioneer and given me a leading position in this particular field. Before the invention of this instrument practically nothing was known about this subject." This was not only Slichter's judgment but also that of Robert Follansbee who wrote a history of the Water Resource Board and stated that this method "aided greatly in changing the character of ground-water investigations to a quantitative basis." There was a considerable demand for Slichter's apparatus. One inquiry came from Sweden and among those sold, one went to Egypt.

It was not only engineers who heard of this. It was reported to Town and Gown on May 10, 1902. The following is what the humanist, Edward T. Owen, minuteman of the evening, professed to have gathered from the discussion:

~ . . . *The flow of this water exercised our speaker's ingenuity. He makes a pair of holes in the ground, puts some nasty stuff in the water of one, and makes his attendant drink from time to time the water of the other. Noting the expression of his face, he determines the moment when the water of one hole appears in the other, and ultimately its rate of progress—some 8–20 feet—or was it miles?—per diem.*

At one place he noted that the surface had sunk some 50 feet below the surrounding level. Reasoning from this phenomenon, he cleverly deduced the resultant existence of a hole, triumphantly downing all skepticism by showing a photograph of the very hole itself. Great are the achievements of Science!

In 1905 Slichter published No. 140 in the *Water Supply and Irrigation Papers of the United States Geological Service.*

No. 140, "Field Measurements of the Rate of Movement of Underground Waters," must have been of considerable importance to those working in the field. He used a number of examples in, for instance, Long Island (dealing with water I was then drinking as a young boy in Brooklyn), California, Kansas, Nebraska, and Texas, to illustrate his methods. He incorporated with only minor changes much of the paper just described concerning the underflow streams of Southern California. This work has many delightful passages, especially those that point out how much more remarkable is the consistency of nature than her occasional temperamental behavior, and how seemingly inconsistent results frequently arise from faulty data or incorrect interpretation.

Two other features of this paper deserve special comment: (1) In it he describes laboratory experiments with both horizontal and vertical tanks (some of them filled with sand from Picnic Point, near Madison, Wisconsin, and some with local glacial deposits) which both corroborated his theoretical treatment and checked the validity of his electrolytic method of measuring direction and velocity of flow of underground waters. (2) This is perhaps the first widely circulated paper that marked Slichter's growing interest in the economics of water supply. In it he discussed the cost of reclaiming a given amount of water, and the proportion of the water pumped which becomes effectively usable.

Slichter's paper was transmitted to the Director of the Survey by Newell with the following comment:

~ *This paper should be interesting and valuable to engineers and geologists, and the direct application of the results to the study of* problems of vital interest to the users of artesian waters should be of great practical value to the general public. A very suggestive and interesting description of the California method of sinking "stovepipe" wells deserves the attention of drillers in unconsolidated deposits throughout our country, while the description of the carefully made tests on typical pumping plants in Texas and New Mexico should appeal to engineers and others who are interested in the problem of raising water for irrigation or other purposes.

If we consider this sequence of publications, and others not listed, from 1899 to 1905, we find a remarkable combination of accomplishments: (1) the development of a mathematical theory of the flow of underground water, (2) the checking of this theory by both field and laboratory measurements made by methods either developed or improved by Slichter, (3) the carrying of instrumentation and theory into the field through work done under Slichter's direction, in which he participated in the actual physical labor. In this third stage (described in the section on Slichter as a consulting and managing engineer) much attention was given to the economic aspect of the subject.

Other Scientific Work. Perhaps the scientific topic which, after underground water, received the most attention from Slichter was the interrelations existing among the density of material within the earth, the pressures existing therein, the rotation period of the earth, and its ellipticity during various geological periods.

In 1898 he published a note in the *Journal of Geology* on the "Pressure Within the Earth"; and in 1909 he wrote a small section of a volume published by the Carnegie Institution of Washington on *The Tidal and Other Problems,* the other papers being by T. C. Chamberlin, F. R. Moulton, W. D. MacMillan, Arthur C. Lunn, and Julius Stieglitz, all of the University of Chicago. Slichter's contribution was entitled: "The Rotation-Period of a Heterogeneous Spheroid."

Probably the most important conclusion of the 1898 paper is that, under rather broad assumptions, "The pressure at the center of a heterogeneous spheroid differs from the pressure at the center of the same matter in the spherical form, by a fractional amount which is less than two-thirds the ellipticity of the spheroid." Thus, with the small ellipticities considered, the use of a theory for the sphere yields good approximations of the pressures involved.

Slichter's feeling for the drama of nature is shown in the final passage of this paper:

~ . . . It must be remembered that there are causes at work which may augment the effects of a change in the internal pressure, and may even produce large results from what seem to be small causes. For example, if we suppose a contest in a given region in the interior between extreme heat on the one hand, and extreme pressure on the other hand, as to whether the material, or a single constitutent of the material, will take on the crystalline form or not, we have a case in point. It may happen that a very slight increase in pressure may materially extend the zone in which crystallization may take place and thus result in a considerable increase in density; it is not impossible to believe that such a zone may exist in the region near the center, where pressures may be dominant on account of their enormous magnitude, and also in a region near the surface, where pressure may again be dominant owing to the lower temperature.

In the first paper he assumed that the moon and the earth had once been one, whereas the second paper appeared in a volume largely concerned with a discussion of the Chamberlin-Moulton planetesimal hypothesis.

The relationship of Slichter and Chamberlin was one of mutual admiration. Thus, on December 10, 1901, Chamberlin (then at the University of Chicago) wrote Slichter, "Your computations . . . constitute a valuable contribution to the discussion of fundamental geological questions. Besides their general value, they would have to me a special one in that, unless on further study I see myself in error, I should like to found on them an argument against the widely accepted proposition that the earth has changed its rate of rotation in any notable degree during its known history."

A little over a year later, January 10, 1903, Chamberlin told of a grant from the Carnegie Institution and wrote: ". . . A portion of the grant is to be used to secure the cooperation of those who are specially competent to attack certain of the problems involved, and I naturally turn to you for cooperation relative to problems connected with the effects of supposed changes of rotation of the earth, as you have already studied this problem."

As was to be expected, Slichter's praise was more expansive. (He never damned with faint praise nor, for that matter, indulged in faint damns.) On November 23, 1907, when the book on the tidal problem was in preparation, he wrote Chamberlin:

~ I have gone over with a great deal of care your very remarkable paper, and must express to you the great pleasure it has been to me to *follow the reasoning which you have brought to bear from so many fields of thought upon this fundamental problem. I know of no piece of work in science that has appealed to me so strongly. It goes back to the fundamental methods of the great masters like Newton and Laplace. I am thoroly convinced that the method of attack that you have adopted is much better than any method dependent upon mathematical analysis that can be applied to these problems. I do not know of a stronger piece of naturalistic logic.*

Twenty-one years later he returned to the same theme in acknowledging Chamberlin's book, *The Two Solar Families.*

~ *I am not at all surprised at the power and sanity with which the difficult problems are approached. Whenever reading your contributions to Science I am always impressed with the conviction that the multiple scheme of approach of the naturalist seems to have over the narrower line of mathematical discussion. The distinction between the two methods is fundamental. In the naturalistic method every item is brought into the problem. Mathematical methods of necessity are characterized by the omission of many things in order to simplify conditions so that mathematical analysis may be brought to bear.*

Also on November 23, 1907, he sent to Chamberlin a critique of a draft of MacMillan's paper, "On the Loss of Energy by the Friction of the Tides." An extended quotation from this is justified as showing the rangy quality of Slichter's mind and his quantitative-qualitative first approach to a subject:

~ *To name one instance in which there is a dissipation of energy, not included in the discussion under consideration, only requires me to direct your attention to the dissipation of energy that takes place on account of the motion given to groundwaters by the tidal wave. This motion is known to exist where conditions are suitable thru considerable areas under the ocean beaches and in the lands adjacent to the shore. It would take some time to make an adequate estimate of the area and extent of material in which groundwater motion is affected by the tides. If computation should show this area to be considerable, we would know at once that the conversion of energy into heat would be here an item of some influence, as the friction of water moving in small pores is of a sort best calculated to produce the conversion into heat energy that I mention.*

I also raise the question whether a portion of the tidal energy should not be considered as converted into potential energy at each tide. In order to present my thought in this matter, let us suppose that there are 40,000 miles of ocean shore which possesses a mean tide amplitude of 4 ft., and that along this shore there is a beach or marsh, averaging 100 ft. in width, which is wetted by high tide, and that ordinarily would not be wetted if tides did not exist. On account of this enormous area which receives moisture from the tide, there is exposed to the evaporating influence of the sun an area which would not be exposed if the tides did not exist. Let us assume that from this area the rate of evaporation is 6 ft. per annum or $\frac{1}{60}$ of a foot per diem. The water that is evaporated is not available for return with the receding waters, and for the moment we will consider it as representing a loss of potential energy. We will assume that the total amount evaporated is lost at an elevation of 2 ft. above its normal position. If the estimate be made on this basis it will be determined that the conversion of energy into potential energy will amount to about 500,000 foot-pounds per second. The evaporation loss at low tide is no more than $\frac{1}{4}$ of the total loss above estimated, so that the deduction on this account is not enough to modify the order of magnitude of the result. We must assume that this water returns in the form of rain, but the return is not at high tide nor mean high tide; the return is at mean sea level. On this basis we may show that, on account of the tide, the sun and the earth are acting as a heat engine, a portion of the work of which is directing against the rotational energy of the earth.

The only important scientific paper that Slichter published in a mathematical journal was on "The Mixing Effect of Surface Waves." This is a neat piece of applied mathematics. His ability to find applications of mathematical analyses in areas other than those in which they first arose is illustrated by the fact that the methods used to secure results concerning the effect of waves on the temperature gradient in a pond are seen to apply to the spread of contamination in slow-moving ground waters. This arises from the circuitous paths taken by the water particles. He states, "The consideration of the mixing, or spread of contaminations, in groundwaters lead to the statement of a very general problem of much interest, which the writer will consider in another place." But "The best laid plans . . . ," especially as war approaches, and for an incipient dean.

As the foregoing portion of this chapter indicates, Slichter's scientific papers were published mostly between 1898 and 1911. In fact there seems to have been no publication from 1884, when as an undergraduate he had a note in *Science* on "Rotation Experiments on Germinating Plants" (he grew seedlings in a rotating dish, showing that the roots tended to grow in the direction of the resultant of the centrifugal and gravitational pulls, and the stems in the opposite direction), until 1898, when he published the paper on "Pressure Within the Earth."

These facts may reflect the fashion of the times more than any peculiarity of Slichter's career. The above-mentioned 1898 paper was published in Volume 6 in the *Journal of Geology*; thus, at the time, this periodical had been in existence for only six years. During the eighteen eighties and nineties outlets for the slowly growing mathematical activity in this country were meager. There was no "publish or perish," but perhaps one had to write texts or starve, for during this period Slichter had been coauthor with Van Velzer of three mathematical textbooks for engineers and a high school text.

In addition to articles that appeared over his name, he made significant contributions to the work of other people. Instances follow.

Slichter gave important mathematical aid to Babcock in connection with the development of a so-called "Babcock milk test" to determine the percentage of butterfat in milk. His contribution made feasible rapid computation of the results of the test and the determination of the market value of the milk. This was essential to the practical use of the test.

As already described, Slichter's work on underground waters apparently started with the help he gave to Professor King.

In April 1903, William H. Hobbs, at that time Professor of Geology at the University of Wisconsin and later a geologist at the University of Michigan, published in the *Bulletin of the Geological Society of America* a paper on a disk-shaped meteorite found near Algoma, Wisconsin. Slichter wrote an appendix entitled "Discussion of the Motion of a Discoid Meteorite," on the probable rotation and orientation of the meteorite as it passed through the atmosphere. The results from this traumatic experience were consistent with the scars on the surface of the meteorite. The mathematics was similar to the standard discussion of the motion of a top. I presume the paper, with proper alterations, would apply to the game of frisbee.

It is almost amusing how often Slichter's work was connected with rotation—whether of the earth, a centrifuge, a dish of seedlings, or a meteorite. It must have seemed like old times to him when, shortly after his retirement, the University, with the help of the Rockefeller Foundation (initiated by Warren Weaver), secured one of the first Svedberg ultracentrifuges in America.

Summary. Slichter's well-deserved scientific reputation is chiefly based on his distinguished study—both theoretical and practical—of underground waters. He did other significant work in applied mathematics, especially in connection with investigations of the interior of the earth and its geological history. He showed quick understanding of other people's problems and ideas and rendered vital mathematical assistance to such scientists as Babcock, King, Hobbs, and Chamberlin. He made no contributions of importance to pure mathematics, nor is there evidence that he was interested in doing so.

Engineer

It is likely that Slichter would have resented separating his work into categories such as science and engineering. There seemed to be little distinction in his own thoughts between the two, although his theoretical paper of 1899 is somewhat set apart from the other groundwater papers because it was addressed to a smaller clientele. The perception of great natural processes, the fitting of a theorem to a physical phenomenon or the invention of a gadget, all were pleasures. If he rated them at all, it would have been as much by the fun they generated as by the degree of their highbrowism.

The publication of "The Theoretical Investigation of the Motion of Ground Waters" was the first fruit of his work as consulting engineer and the origin of much of his later work of this type. It would not be worth while, and probably not possible, to list all of his engineering and consulting projects, but we can distinguish several types: (1) services rendered in the employ of the Geological Survey, much of it in the Reclamation Service; (2) services to other public agencies, especially in regard to city water supplies; (3) services to private firms or individuals connected with irrigation; and (4) legal disputes.

From the time he was first appointed a hydrographer for the United States Geological Survey in 1897 (when he was granted $400 to help Professor King in his study of underground waters and especially his development of the aspirator for measuring the porosity of sand and gravel) until 1912 (when he left the Reclamation Service), he was busy surveying water resources and assessing their availability throughout the United States, with some overlap beyond both its southern and its northern borders. Thus, starting about 1902, he was active in California, in the Rio Grande Valley, on Long Island, New York, and in the Oklahoma-Kansas-Nebraska region. This employment was on a per diem basis and ranged from making short feasibility reports, to being the engineer in charge of irrigation projects—notably the one near Garden City, Kansas.

In dealing with this bewildering pre-air-travel activity as a consulting engineer for the United States Geological Survey, a rather detailed description will be given of Slichter's work in and near Garden City, Kansas, from 1904 to 1910; and a cursory survey, illuminated here and there by an interesting or amusing item, will suffice in regard to his other engineering activities for the Service. This choice is made for various reasons. The documentation is more complete for Garden City than for the other projects. The Garden City project was one in which he was in charge of every phase—from the exploration of the water resources and their availability to the construction of wells and the arrangement with the farmers who would be served. Moreover, it would be monotonous to report on well after well and gravel after gravel in river valley after river valley.

Garden City Project. Garden City is located in western Kansas, on the Arkansas River, in Finney County. It now has a population of over 15,000, but in 1900, a few years before Slichter was first connected with Garden City, the population was 1,590. It had been considerably greater during the boom days of the 1880s. By 1910 it had almost doubled (3,171). It was and is a station on the main line of the Atchison, Topeka, and Santa Fe Railroad.

In a memorandum entitled "History of Garden City Project" Slichter described the start of this work as follows:

~ *A reconnaissance was begun in June, 1904, in the Arkansas Valley in the vicinity of Garden City, Kansas, in accordance with authority given by the Chief Engineer of the Reclamation Service in a letter dated April 30, 1904, in which specific allotment was made for carrying on the work. The purpose of these investigations was to determine whether or not the underground water could be economically obtained in sufficient quantity for irrigation. The rate of the underflow was measured and the position of the normal water plane determined. A full account of these investigations can be found in the Third Annual Report of the Reclamation Service.*

This gives little idea of the effort that went into this first year of the project. Slichter spent considerable time in the area, but also placed much responsibility on Ray Owen and, later, on Henry C. Wolff.

Ray Owen, who was about twenty-five, was a graduate of the College of Engineering at the University of Wisconsin and, we believe, a former student of Slichter's. Owen was to join the faculty of the University in the field of topographical engineering the next year and thence to rise through the various ranks until he retired in 1949. For many of the remaining eighteen years of his life, after 1949, he did professional surveying in and around Madison. It was a great opportunity for Owen to show his own initiative under a chief who appreciated it and who could write to him in July, "The reports you have made on these wells pleased me very much indeed. I think you have done very thorough work and have obtained exactly the information we desired."

One of Owen's more exciting experiences was when the Arkansas River flooded. He wrote on October 4, 1904:

~ *The river rose yesterday to an unprecedented height for this season of the year. The men were across the river and could not get back and have not yet been able to get across. They are putting wells down along the Perry ditch. The headgates are liable to go at any time now. The river has cut 100′ North into the bank. We have 3 sets of wells under water. The yesterdays mail has not yet gone. This* P.M. *I will get it of the carrier who is at Dick Maphets [?], wade or swim the river and get a horse to go to town with, and take this mail and get the three days of mail which we have coming. If I cannot clear up things here I will have Harsha [Harza] or Taggart to pull wells and ship freight, or take it to Perry's and leave Saturday* A.M. *on schedule time.*

One of our wells, which we fortunately had pulled, was just where the main current of the River is now. There are a two days' collection of teams and cattle waiting at the ford on both sides of the river. Made a megaphone this A.M. *and shouted directions across to Taggart.*

Yours respectfully,
Ray Owen

P.S. If you don't get this you will know I got drowned crossing the river.

Henry C. Wolff, a member of the Department of Mathematics, did much work under Slichter, elsewhere as well as on this project. He had graduated from the University of Wisconsin in 1897 and was a member of the staff of the Department of Mathematics from 1900 to 1919, and, as already stated, got his Ph.D. under Slichter in 1908. There had evidently been some bureaucratic ballup in getting Wolff appointed as an irrigation engineer. Slichter's protest was explicit for, on October 19, he wrote the Civil Service Commissioners:

~ *I have received your telegram of this date signed Cooley, stating that no application had been received from Mr. Henry C. Wolff for admission to the Civil Service Examinations at Madison for irrigation engineer,—examinations to be held on October 19 and 20, 1904. I desire to say that the application to take this examination, together with acompanying information and vouchers, were duly filled out and mailed at Madison Post Office on October 3, 1904. These papers and application were placed in a large envelope addressed "Civil Service Commissioners, Washington, D.C.," with two 2-cent stamps on the outside, and placed in the letter chute opening in the lobby of the Post Office at Madison, Wisconsin, about 7:40 P.M. October 3d. The loss of these papers has caused the very greatest inconvenience to Mr. Wolff and myself and I respectfully suggest that you request that a search be made for the missing papers.*

It would have been fascinating to witness Slichter joust with an IBM accounting system.

A third assistant was Frederik T. Thwaites, then an undergraduate at the University of Wisconsin, who was to be for many years curator of the geology museum at the University. He evidently was homesick, for Slichter wrote Owen suggesting that he read the following to Thwaites to "jolly him up,"

> *Matilda ate cake; Matilda ate jelly;*
> *Matilda went home with a pain in her—*
> *Now don't be excited and don't be misled,*
> *Matilda went home with a pain in her head.*

Thwaites, however, was a collector of ancient things besides rocks, and he informed Owen that the ditty was "antiquated"; nor was Owen a man to allow it to go unmatched, for in an epilogue to a report of August 14, we find,

> *Tom fell down the elevator,*
> *Wasn't found till three months later.*
> *All the neighbors said "Gee whizz!*
> *What a spoiled boy Tommy is!"*

We also have from 1904 an interesting list of groceries which Slichter had kept, although it is not in any handwriting that I have come to recognize. The substitutions seem to be more by price than by purpose, hence four pounds of coffee at $1.20 instead of fifty pounds of flour at $1.25, or five pounds of strawberries at 40 cents instead of one sack of cornmeal at the same price. On the whole it is a masculine list, even to the question mark opposite "Gold Dust." This list is reproduced on the opposite page.

Slichter wrote two letters in the summer and fall of 1904 to Wesley Merritt, Industrial Commissioner of the Santa Fe Railroad, which in-

Garden City, Kansas, ——July 1/04'   190__.

S. Geological Surveying C,

In Account With CARTER, WARNER & CO.,

HARDWARE, GROCERIES, LUMBER AND FARM MACHINERY.

GASOLINE ENGINES AND WIND MILLS A SPECIALTY.

June	13	To 1 Gal. Coal Oil, 25 / 2 Doz. Lemons 50		73
	15	50# Flour 1.25 / 4# Butter 80 / 3 Cans Sardines 40		2.45
	"	3 Cans Tomatoes 55 / 5 Gal. Gass 1.25		1.80
	16	2 Doz. Lemons 50 / Vanilla Ex. 20 / Lemon Ex. 10		80
		Cheese2# 30 / 10# Sugar 65 / 1 Pack. Force 15		1.10
		2 Packages Oats / 8 root leaf nut 1.00 / 5# straw berries 40		2.0
	18	5 Gal. Gasoline		1.25
	20	16# Sugar 1.00 / Matches 15 / Pack. Gold Dust 25		1.40
		1 Package Oat Meal 25 / 1 Sack Corn Meal 40		65
	21	1 Doz. Lemons 25 / 14# Ham 2.35 / 2 Doz. Lemons 50		3.10
		5 Gal. Gas. 1.25 / 1/2# Tea 25 / 1 Cake Turner 10		1.60
		1 Cake Griddle 25 / 4 Cans Tomatoes 50/ 1 Kettle 1.00		1.75
	23	16# Sugar 1.00 / 2 Packages Oat Meal 50		1.50
		1 Peck Potatoes 40 / 2 1/2# Cheese 35		75
		1# Bak. Powder 50 / 1 Cream Can. 35		85
	25	1 Pack. Gold Dust 25 / 3# Coffee 1.00 / 2 Cans Pumpkin 25		1.50
		5# Lard 75 / 51/2# Beans 35 / 1 Pack. Soda 10		1.20
		3 Pack. Macoroni 30 / Gloves 10		40
		1# Cocoa 25 / 1/2# Chocolate 25 / 6-1/2# Bacon 95		1.45
		1 Oil Bbl. / 8 root leaf nut 1.25		50
	27	50# Flour 1.25 / 16# Sugar 1.00 / 1 Doz. Lemons 25		2.50
		1 Package Yeast 05 / 2 Box Tacks 10 / 1 Broom 30		45
	28	1 Bush Potatoes 1.60 / 2 Doz Eggs 30 / 2 Doz. Lemons 50		2.40
	29	2 Boxes Berries 20 / 1 Hasp 10		30
				30.63

dicate his geological findings and his intense concern for the financial
aspect of the irrigation projects in the area. The first of these was on
August 2; excerpts follow:

~ I find that the multitude of the underflow is very much greater
than I had anticipated. The facts of the case seem to be about as

follows. The underground drainage in this region is so enormous and the freedom of passage of water through the gravels of the underflow is so great that there is no surplus water left to form surface streams or to form a perennial supply for the Arkansas River. The average rainfall for this very narrow region is about 20 inches per year, a large part of which passes into the porous soil, so that the actual total amount contributed to the underflow must be considerable. In any ordinary region where the gravels are but a few feet or a few score feet in depth this rainfall would supply perennial streams. But the gravels beneath the plains are so coarse and so deep and extensive in width that the seepage waters are not obliged to seek release in surface streams, but the underground conditions are such that ample drainage is afforded to the entire area by means of the porous beds of gravel beneath the surface. We have found an almost uniform rate of movement of this groundwater in all parts of the valley, varying from 9 to 11 feet per day and the direction of motion almost exactly due east in line with the general surface drainage. In certain places we find strata of extremely coarse gravel in which the velocity of the underflow represents a maximum of about 24 feet per day. . . .

I know of no place in the country where groundwaters can be obtained at so little expense for well construction and in such enormous quantities. Even at the high price of gasoline that prevails in the valley, the cost of pumping with gasoline engines is not high. All that is necessary to develop this country is to have these facts generally understood.

By the end of September he was struggling with the price of fuel and wrote to Merritt concerning a large well owned by a Mr. Holcomb which needed coal at $3.50 or $4.00 a ton rather than the prevailing price of $4.50. He also discussed the fuel problems connected with small wells.

~ In a paper which I put in your hands a few days ago, I called your attention to the cheapness with which water can be recovered near Garden City by small pumping plants, if crude oil generators are attached to the gasoline engines at present in use. By far the larger part of the pumping in the neighborhod of Garden City will always be done by small plants covering from 75 to 150 acres. For such plants, the ideal fuel at present is crude oil used in crude oil generators attached to the gasoline engines. The discovery of oil in Kansas gives the possibility of an ample supply of crude oil for such purposes in the Arkansas valley. If the settlers begin to take up the use of crude oil, as I have advised, it will be important to secure railroad rates on crude oil in carload lots from Chanute to the western Arkansas valley. I suppose, at present, there is no rate at all. I find the price of crude oil at El Paso, Texas, which is 800 miles from Beaumont, to be 3¢ per gallon in large lots. Garden City is 350 miles from Chanute, and it would be the greatest thing that could possibly happen for that western country, to be able to get Chanute crude oil at the price of about 3¢ per gallon. This would solve the problem of fuel oil for small plants very satisfactorily indeed, and enable them with limited capital to be sure of a successful career in irrigation in the Arkansas valley.

Oil wells are now within a few miles of Garden City but the irrigation pumps use natural gas from the immense underlying field. Perhaps it would have been fortunate if Slichter had stuck to promoting irrigation from individual wells.

By now Slichter had a full head of steam, and on November 8 wrote Owen: "I shall leave Kansas City on #7 on the afternoon of November 12 or 13, exact date not yet certain. I shall probably wire you from some point on the route to meet me on the train at Garden City and ride with me as far as Syracuse. You will be able to return on #7 in the morning. This will get me up at 2:00 o'clock in the morning, but I hope that the train will be late so as to make the time more convenient." He did not seem worried about Owen's sleep; apparently no permanent harm was inflicted since Owen lived to be eighty-nine. Owen must have listened rapidly, for the run of the fifty-two miles from Garden City to Syracuse took only one hundred minutes. Another problem is presented in that #7 did not run both east and west —a trivial matter between engineers.

The year 1904 had been used chiefly in assessing the underground water resources of the Garden City area and in studying their availability; and 1905 in making plans for a pumping plant to greatly increase the supply of water for the Farmers' Ditch, an already constructed artery.

In his history of the project Slichter described the 1905 campaign as follows:

~ Full report on conditions found was submitted to Washington on March 8, 1905. At the same time an outline description of a proposed pumping plant for the reclamation of land near Deerfield, Kansas, was submitted, together with estimate of cost of construction and operation and estimate of cost of an infiltration gallery. A Board of Engineers, U.S.R.S., met in Denver March 24, 1905, to go over the proposed plans. This board recommended that the surveys and investigations be continued in order to obtain more definite information affecting probable cost of construction and operation and the extent and location of lands proposed to be irrigated, and that alternative estimates of several possible designs and limiting sizes be prepared for future consideration. These further investigations were approved by the Chief Engineer under date of April 7, 1905, and authority given in a letter dated April 18, 1905, for carrying on the work.

During the field season of 1905 the surveys and investigations were continued. On June 7, 1905, the underflow of Arkansas River at Deerfield, Kansas, to the amount of 200 second-feet was appropriated by the Reclamation Service and notices were duly filed and posted as required by the laws of Kansas.

Final plans and estimates were drawn up and submitted to a Board of Engineers which met at Garden City, Kans. on Sept. 5, 1905.

Slichter was not acting just as a modern water witch and prospective construction engineer. He also was deeply involved in the financial feasibility of the plan. This is illustrated in his letter of February 27, 1905, to W. M. Bell, Secretary of the Finney County Farmers' Irrigation Association. Since this letter gives a clear picture of the whole undertaking it follows in full:

~ I have your letter of Feb. 22. In reply will state that I have designed an irrigation project, under the Reclamation Act, which contemplates the recovery of one hundred second-feet of groundwater near Deerfield and placing same in the Farmers' Ditch. The plans for this plant are now about complete, and the Chief Engineer of the Reclamation Service has called a meeting of a Board of Engineers to meet at Garden on March 27, to pass upon the estimates and plans submitted by me.

If the action of the Board of Engineers should be favorable, the matter will then go before the owners of lands along the Farmers' Ditch. If they desire to have the proposed works constructed, they will have to vote to that effect; and if the works are constructed, the payment for same will extend over a period of 10 years, payments to be made in ten equal annual installments, without interest. A full explanation of the project will be made at the time the board meets.

The Board of Engineers will want to look over the ground, and it will be advisable for those of you that are especially interested in the matter, to be on hand at that time to assist as far as practicable in making known your wants, and to assist in giving the engineers accurate information concerning the country tributary to Garden and its future development.

Perhaps I ought not to say anything in regard to the project in advance of the meeting of the engineers, as they may not coincide with my views. I believe, however, that my plans constitute a practical means of recovering one hundred second-feet of groundwater and placing same in the Farmers' Ditch. I believe that this will give a very much more reliable source of supply than exists anywhere along the Arkansas River, even in the State of Colorado.

The plant I propose consists of a large installation of pumps and conduits, the cost of which will be between $200,000 and $250,000. This plant will surely recover, during a season of 150 days, about 30,000 acre feet of water. This is sufficient to irrigate from 10 to 15 thousand acres of land. The running expense of such a plant will amount to about $11,000 a year for fuel and labor and miscellaneous supplies; and about $13,000 a year additional for depreciation on the plant. The figures that I offer for depreciation are made excessively high, and my estimates of cost I have attempted to make as high as practicable, in order that the estimates may present the matter in the most unfavorable light possible.

Including all of the expenses mentioned above, the total cost of water will amount to 80¢ per acre covered one foot in depth. If the initial cost of the plant should run as high as $250,000, this would be a maximum cost of $25 per acre for the land reclaimed, which would require an annual payment of $2.50 per acre for a period of ten years. My faith in the quality of the land near Garden, under the Farmers' Ditch, is such that I believe it is worth much more than $25 an acre to secure a reliable supply of water.

Slichter was also busy answering questions concerning the terms on which water would be furnished to the farmers. Not only was the price important but also the scheduling of the delivering of water. In a letter of August 1, a water user, P. de Pronleroy, inquires and remarks:

~ Shall we have to take turns as now, regardless of needs or will we be free to choose our time according to condition of crops?

The value of water lies in the possibility to apply it at the right time, for then it makes a crop. Otherwise it would have the same value as rain, would depend on "a chance" of getting it at the right time.

I'm not sure that Slichter's reply removed all doubt but it probably was as much reassurance as he was authorized to give:

~ . . . It is the plan in all government projects to provide a sufficient amount of water so that no one will suffer for its need at any time. This, however, might not mean that every individual will be able to get water just whenever he happens to telephone for it, for it is obvious that it would not be the best management to operate all the laterals on the ditch during all the time. This matter would have to be arranged, however, so there would be no opportunity for any suffering from lack of water. At the same time it would not be possible to arrange matters so that the delivery of water could be postponed at a certain point just because the owner of the land had some other engagement on that particular day. Provision should be made so that each individual can get the water at very frequent intervals so as to encourage the growing of high grade crops and securing the very best yield under all circumstances.

Slichter's letters dealing with personnel show fine concern for their welfare. Thus, on June 20, he wrote to the pastor of the Presbyterian Church at Garden City: "A young lady, who will arrive with me at Garden City on June 23, 1905, and who holds a position as stenographer in the United States Reclamation Service, will desire to secure a room in Garden City for several months. I will call upon you when I arrive at Garden City, and I trust that you will know of some place where she would have a pleasant home." On July 8 he wrote to John H. Gabriel, a Denver attorney, a graduate of the University of Wisconsin, concerning a young assistant:

~ I am sending to Denver one of my engineers, Mr. B. F. Harza, who has had a very painful experience the past two or three weeks with facial neuralgia, and he goes to Denver to have an operation performed to relieve him. I am sending him to you with confidence that you will recommend to him a first class surgeon or physician who can take him to a good hospital where the operation can be performed. The operation is a very simple one, but, nevertheless, I am anxious that Mr. Harza receive the very best care.

Mr. Harza is a university boy, at present in the senior class in the Civil Engineering Course.

Mr. Harza has come in from camp late this afternoon, after the banks have closed, and I am unable to furnish him with the cash for the expense of the operation, but I have given him a check against my account on the Garden City National Bank, which I hope you can aid him in cashing at Denver. I know you will take an interest in Mr. Harza as a University boy, and I thank you very much for any attention that you can give him.

Gabriel willingly helped. Harza became a distinguished engineer and many years later, through the firm of Mason, Slichter, and Hay, successfully located leaks in dams, using the device developed by Max Mason and others, including Slichter's son Louis (see Chapter 6), for detecting submarines.

Evidently Slichter's son Sumner, then aged twelve, had visited Garden City, for on August 17, 1905, Slichter wrote a typical family letter:

~ *Sumner Huber Slichter, Esq.,*
   636 Francis St.,
      Madison, Wis.
Dear Sumner,

   I have your letter of Aug. 15th. I will send you some of your clothes by mail. I think you had better wait for your comb and brush until I get back.

   We are having a nice time at Garden City. Yesterday we had a picnic at the Holcomb Ranch and had all kinds of canned Chili to put on our sandwiches and pie. We also drove out on the uplands north of Garden to look at the land that is under the Farmers' Ditch. We saw some very nice beets, but we saw more places where there ought to be beets than we did where there were beets.

   Papa will be very glad if you will go down town and look at the new building; then write me how they are getting along with the work, especially whether they have the tar roof on, and whether they are getting ready to plaster the store on the first floor.

With love

Nice as this is, I like even better the letter to which it was a reply— a letter which was written under the letterhead of the Reclamation Service, an early example of Sumner's lifelong habit of writing to his father on snitched stationery.

~ *Dear Papa:—I received your letter. I hope Garden City gets the sugar factory. Please send me my comb and brush, linen pants and if they are out there a pair of blue pants. Grandpa celebrated his birthday last Saturday. It was on Friday but he put it off a day so that I could be present. We had tomatoes, ice cream, a birthday cake, plain and whole wheat bread, cold tongue, fine jellie, and cookies. Donald Byrne and all us boys were there. It was fine. We gave him some wine. Mama was there but did not stay for supper.*

With love
Sumner

   The culmination of the 1905 work was the authorization for the construction of a plant at Deerfield. One of the warmest congratulations which Slichter received was from Victor Murdock, who for many years was editor of *The Eagle* of Wichita, Kansas, and in 1905, at the age of thirty-four, was a representative to Congress, where he remained until 1914. His district at the time included Finney County. He was a participant in the Bull Moose Movement of 1912. He could match Slichter for ebullient enthusiasm and on October 12 he wrote:

~ *My dear Mr. Slichter.*

I take this occasion upon behalf of all my constituents and for myself to thank you for the development of the Garden City project. I realize, more completely than others I think, how far a step you took in advance when you recommended it. I began my study of the subject as pertaining to western Kansas with Johnston's "The Utilization of the High Plains." You remember, of course, his discouraging and disheartening verdict. I knew that some day he would be reversed, and the evening you talked with me in the Cochran Hotel, Washington, I knew that you would be the man to reverse him. A week ago I thought, for a moment, that we might lose all. But my suspense was only 48 hours long. A wire came back finally that the Secretary had approved, and that the perilous recall was out of the way. On anticipation alone, a new era has set in out on the prairies, and we all realize that another day is dawning on our part of the slope.

You talked with me upon another subject which I have not neglected—the production of untaxed methylated alcohol for fuel. After a study of the subject, I tried to put through the House, a preliminary resolution but, after a favorable committee report, the guardians of the revenue, smashed it. But it must come. Germany, France, the Netherlands are all advancing along that line. Even little Venezuela is making some progress. Any literature you may know of upon the subject, I would be glad to have you cite me, as I intend to keep at it.

The *Garden City Herald* of June 2, 1906, had large headlines—"WATERING THE DESERT PLACES," with a subheadline—"The Greatest Irrigation Project of the United States Government about to be Installed."

The article which followed said: ". . . Through the persistent and faithful work of Congressman Victor Murdock Garden City was selected as the place for making the first experiment." And later: "Professor Slichter of the United States Reclamation Service and a force of expert engineers were sent here to investigate the water supply and report upon the feasibility of a pumping plant." The paper reported that this led to a unanimous favorable decision of the engineers.

A few years later, in connection with a dispute between Kansas and

Colorado over the use of water, the *Garden City Imprint* reported: "The Arkansas River is what is known as an upside down stream. That is, it is a stream where the greatest part of the water flows in the sand and gravel underneath the surface and is therefore not visible to the eye."

The project which Slichter had proposed and was to supervise consisted of a series of wells on both sides of the Arkansas River, situated in such directions and distances from each other as not to interfere with the supply of water to any of the wells. The water from the south side of the river was to be siphoned (and pumped) under the river and, with water from the north side, placed in the Farmer's Ditch, which, although already established, often got insufficient quantities of water from the river. This involved in all a ditch, about twenty wells with their small individual pumps, a power plant, and a pumping station.

Murdock was not the only person in Congress approached by Slichter on the subject of the alcohol tax, for in 1906 he wrote to Senator John C. Spooner of Wisconsin, one of the few men who received two honorary degrees from the University of Wisconsin.

~ *I am greatly interested in a bill now before the Senate for the removal of the internal revenue tax from denaturalized alcohol for use in the arts. I have personal knowledge of the value that such an act would be to many communities in the far west which are removed from the petroleum markets, and which at the same time have at hand material for the manufacture of alcohol. This is especially true in the western communities that have beet sugar factories. It is difficult at the present time to find a suitable use and market for the crude molasses that results from the manufacture of the sugar. At the same time it would be very easy, under the proposed act, to convert the crude molasses into alcohol and use the same in the vicinity of the works in alcohol engines for furnishing power for the pumping of water for irrigation.*

There are also many communities where potatoes are grown in large quantities and some years it is exceedingly difficult to find a market at suitable prices for the product. This is now the case in a large section of the State of Wisconsin. When the price of potatoes is unsatisfactory it will be possible to convert the potatoes into alcohol, and hence avoid the great waste that now takes place in some years in Wisconsin.

I call your attention to the fact that there are several beet sugar factories in Wisconsin that would be benefited by the passage of this bill. In its practical effects the new legislation amounts to the creation of a new industry in this country.

I wonder if someday, when the oil supply has dwindled, the alcohol project may assume greater importance.

Slichter had had two wonderful summers, 1904 and 1905; and although he was by no means through with the Garden City–Deerfield project, it had entered a new phase where the ratio of annoyance to excitement increased. The aforementioned history apparently was written shortly after September 1, 1907, the last date for which facts are given, though there are forecasts for the end of the year. For the years 1906 and 1907, the history is full of details of costs, dates of construction and of the parallel legal work of completing contracts for the use of water, the acquisition of land, and the establishment of rights of way. When bids were too high, direct construction was resorted to.

The details would be of little interest at present; however, they do show the versatility of Slichter's managerial abilities, and the costs that they reveal would now seem unbelievably low.

Slichter as Supervising Engineer was in charge, in an overall sense, of this phase of the work; but instead of working with his own university colleagues and students, he had the collaboration of various consultants, and there was a resident engineer, Charles E. Gordon, in immediate control.

Private bureaucrats could annoy Slichter as much as those connected with the Civil Service. Thus he complained to the general passenger agent of the Santa Fe Railroad on October 17, 1905, that the agent at Garden City had not properly forwarded a mileage book and on three occasions had failed to ship promptly material being sent by the railroad. He was therefore prepared for a letter from Gordon, dated December 17, 1906, reporting that the transportation system was very bad and hence, since he could not get delivery of coal belonging to the government, he had seized, promisng to pay later, coal which the agent said belonged to the railroad. Slichter replied at once:

~ I note in the last paragraph of your letter that you seized a car load of coal stating that a supply was vital for the continuance of the work. I highly approve of this course and have written to Washington quoting your letter and informing the Chief Engineer that I believe that an emergency existed which justified your act, and that I have commended you for your foresight.

Gordon stayed at Garden City and Deerfield until the fall of 1907. He seems to have been very successful in the construction work but somewhat less so in caring for contracts with landowners and water users. Gordon was not the last engineer to have trouble with the users of water from the Farmers' Ditch.

In August 1907, money for the work was running low, and Slichter was given explicit directions to skeletonize the force. This meant that Gordon was to devote himself to finishing the securing of contracts with farmers before November and, meanwhile, turn over the running of the project and construction to C. E. Hogle, an electrical assistant.

Poor Hogle! On August 19 Slichter had informed Gordon that, in accordance with the wishes of Newell, now Director of the Reclamation Service, "Laborers and other men now employed at $2.25 [per diem] and higher, can gradually be let out and new men secured at $1.75 to $2.00. The labor market is in decidedly better condition than a few months ago." (Engineers identify with management.)

At the end of August, Hogle writes of the "chaotic condition of affairs." Staff members were leaving. The inspector who was to stay changed his mind. The next one stayed only half a day. The stenographer was leaving because his sister was to be married. The rodman had left and the fiscal agent had not arrived. However, Hogle had set up notices in the post office seeking laborers at the reduced rates. Luke 14: 16–24 tells the same story more succinctly but scarcely more vividly.

Some of the details that were entailed in a project such as that at Garden City and Deerfield can be gathered from the following: leave for Ray Owen (1905), mortgages on land owned by farmers of Finney County Water Users Association (1906), check valves versus flap valves (1906), recommendation to try *Bromus Imeres* as a grass suitable for holding the embankments of irrigation ditches (1906), method of reporting work-time on government forms (1908), the unfortunate publication of an article concerning the project (1909). The reputation that Slichter had for scorning details must have been based more on the explosive nature of his remarks concerning them than on lack of attention to them.

The immediate success of the project was blocked by two uncontrollable events, and another that was not controlled: (1) the panic of 1907; (2) several years when the rainfall in western Kansas was considerably more than usual; and (3) seepage from the irrigation ditches which was unexpectedly great. Perhaps we should add to this list a certain amount of human cussedness on the part of both farmers and officials.

In October 1907, after some lesser bank failures, the Knickerbocker Trust Company of New York closed its doors. Many banks were immediately in trouble and some collapsed. A few months later, under the leadership of J. Pierpont Morgan, order was reestablished. The panic was severe, but the time of recovery was relatively short. However, no wound is without a period of healing. One of the effects of the crash was the sharp curtailment of government expenditures. This came after most of the construction at Deerfield was complete but undoubtedly slowed down further development.

In the early 1890s and again, centering around 1901, there had been severe droughts in western Kansas. When Slichter started his surveys in 1904, there were vivid memories of those years. But 1906, 1907, and 1909 had rainfalls at Garden City ranging from 11% to 40% above normal, and in the months of June and July, nearly 50% above. The annual rainfall for 1908 was below average, but even for that year the summer precipitation was slightly greater than usual; and at the end of 1910 Newell wrote Slichter welcoming him home from a year abroad and assuring him that, when drought again hit, the farmers would be eager for water. But at the moment they were in no mood to pay for water. It would have been difficult to sell water rights to Noah —this is what economists call the "law of supply and demand." In 1911 there was another drought.

Evidently much more water than expected was lost through seepage from the irrigation ditches, and thus the water that could be delivered to the customers was reduced. Slichter declared that some of the ditches would have to be lined with concrete to make them usable. Naturally the chief worry was the main canal, of which Slichter wrote to Newell on December 7, 1910:

~ The very discouraging reports I last received concerning the behavior of the main canal raise the main question whether it would be best to increase the capacity of the plant if this loss cannot be cut down. I understand from Mr. Hogle that the loss of the water amounted to 40% or 50% in the main canal. If this loss can be kept down to a reasonable amount, it seems to me in every way feasible to enlarge the plant; during the dry years which are sure to come the demand for water in the Arkansas Valley will be enormous, their only source during these years will be the ground waters in the bottom lands of the valley unless the past twenty-five years record is no indication of what may be the case in the future.

Other troubles required Slichter's attention. It was proposed to sink private wells so close to the government wells as to interfere with their water supply. In September 1908, he took this matter up with the Director of the Reclamation Service. After a short description of the proposed wells, he continued:

~ The development above proposed is so planned as to not only very materially interfere with the working of the government pumping plant at Deerfield, but in such a manner that the present plant of the Government will be rendered practically worthless for the purposes intended. Any extensive withdrawal of underground water in the neighborhood of the Government wells will diminish correspondingly the amount that can be drawn from these wells by the Government pumps and will depreciate and render worthless the works at present installed. Some of the pumping plants and test wells of the proposed development by the sugar company lie within a few hundred feet of the Government wells. I cannot make too emphatic the fact that any development of underground waters within three miles of the present line of the Government wells will most materially and vitally affect the yield of the same. This is especially important in as much as the best underflow waters are found on the south side of the river and the best government pumping plants and best wells are the ones that will be primarily affected. . . .

I urge in a most emphatic manner that steps be taken at once to protect the rights of the Government in the underground waters at Deerfield. I expect considerable difficulty in convincing all parties concerned of the serious consequences that will follow this proposed development. My knowledge of the situation at Deerfield is the result of many years of investigation and study and I am absolutely certain of the results that will follow if the proposed development of the sugar company is not stopped immediately.

I believe it possible for the sugar company to devise a proper scheme of irrigating most of their lands by underground waters, and yet keep away a distance of at least three miles from our line of wells. I therefore believe all work of development of the kind now contemplated should be stopped at once, and that any development of underground waters by the sugar company should be so planned as not to interfere with the right of the Government in the underflow waters already appropriated.

We have gone too far not to finish the story, although Slichter had little to do with its final phase. In 1908 the farmers, through the Finney County Water Users Association (successors to The Finney County Farmers' Irrigation Association), refused to pay the assessments, stating that: "On account of the failure of the contractors for the Reclamation Service to deliver the pumps in time for the irrigation season of 1908, the members of the Finney County Water Users Association find themselves unable to meet the charges for construction and maintenance due in December, 1908." They also would not agree to the high assessments for 1909. The Association wrote the Department of Interior, giving a summary of their point of view in a letter of October 18, 1909. The Association claimed that they had been led to believe that the water might cost from three to four dollars per acre per year and that they were being asked to pay about thirteen dollars per acre per year. The letter of protest added, however:

~ In an experimental way the Project has been of vast benefit to the people of this valley and to the country at large as a demonstration of what may be accomplished in recovering underground water on a large scale, but the cost as now charged to the individual farmers under the Garden City Project is excessive when the benefits derived by them is taken into consideration, and relief should be afforded so the cost of the works and the maintenance charge should not exceed the cost per acre as originally understood by the farmers.

The supervising engineer, I. W. McConnell, saw things differently. He had written to the Association on March 21, 1909, complaining of the condition in which the Farmers' Ditch and its laterals were kept. "The entire system has fallen into a condition of disrepair which cannot be described in a satisfactory way as a basis for repair operations. . . . The system is almost entirely devoid of devices whereby any man can assure himself or others that such waters as may be available can be equitably divided among the several water users."
He added:

~ It appears now that the plan did not receive the sympathetic support of the Board; that at its inception the parties charged with carrying out the plan were hampered by diverse instructions from the various members of the Board; and that if the plan was not actually retarded by the active interference of the ditch superintendent, it was at least crippled by his indifference. . . . Under a system which gives no control of the water as to division, and which compels the competent irrigator to bear the burden of the careless and the incompetent irrigator, responsibility for failure other than those occurring at the government plant cannot be accepted.

The Reclamation Service offered to increase the capacity of the plant, but the users refused to agree to additional assessments. The plant was shut down in 1909 and liens placed against the lands of the members of the Association. The government finally sold the plant to the sugar company, its erstwhile competitor for the underlying water. Since then, all the buildings except an office-residence house have burned.

It was not long before the Water Users Association began a campaign to have the liens removed. On August 10, 1913, the Board of the Association endorsed a bill to relieve the farmers of any liability for the assessments of the Garden City Project. The final paragraphs of this endorsement read:

~ *That the Secretary of the Interior levied $6.50 per acre on the lands involved for the season of 1909, which seemed to be confiscatory, no part of which has been paid.*

That said pumping plant has not been operated or attempted since the year 1909, and the lien on the lands under said project continue to be a cloud and menace to future development, and for that reason, the undersigned members of the Board of The Finney County Water User's Association, speaking for all stock holders as well as ourselves, do most earnestly ask that said Bill be passed and the lands relieved of the lien.

It took a little while to get what they wanted; but on June 5, 1920, the following bill was approved, bringing to an end the story of the Garden City Project:

~ *Be it enacted by the Senate and the House of Representatives of the United States of America in Congress assembled, that the contracts affecting land in the Garden City Project, of the Reclamation Service in Finney County, Kansas, heretofore entered into between the Finney County Water Users Association of Finney County, Kansas, or with individual landowners, and the Secretary of the Interior for the supply and use of water from the irrigation plant of the United States be, and the same are hereby, cancelled and released; and the liens upon the lands in said county created by such contracts are hereby released and discharged.*

I visited Garden City in the fall of 1970 and found a prosperous land, the prosperity being based on irrigation. I was told that ditches carrying river water, the Farmers' Ditch among them, provide perhaps a quarter to a third of the water. Individually owned wells with gas-powered pumps provide the rest. A few men, boys in 1904 to 1910, remember Slichter. Although the Garden City Project per se, for an intricate complex of reasons, turned out to be an expensive debacle, the people who know the history are grateful to Slichter. He gave them faith in their land. He made clear what great water resources were available beneath the soil. He helped teach them how to construct and locate wells. His enthusiastic optimism, which may have contributed to the failure of the Project, certainly contributed to the success of irrigation farming. Slichter's work was one of the building blocks of the region's wealth.

Of course I found only a few in the area who knew who Slichter was. This episode is a fine example of how in time a man's material work may crumble, his name be forgotten, and yet his influence continue to be great.

Other Surveys for the United States Reclamation Service. We have already mentioned Slichter's work in California, where in 1902 he first made extensive use of his new method for measuring the velocity of the flow of underground water.

On the great plains, his explorations (some for private concerns) allowed him to roam over the valleys of the Rio Grande in Texas, of the Cimarron in Oklahoma and Kansas, of the Arkansas in Kansas, and of the South Platte in Colorado and Nebraska.

In the last of these, Slichter placed Wolff in charge of much of the work. A large share of the information was derived from already existing wells, sometimes in operation and sometimes closed down. Occasionally, in connection with trial runs, these wells were repaired and cleaned in order to secure reliable results, incidentally benefiting the owners.

On July 3, 1905, he sent Wolff a memorandum notable for the amount and kind of directions and information that he gave. At each of four places—Sterling, Colorado; and in Nebraska, Ogalalla, North Platte, and Birdwood Springs (the last not a town)—he outlined the tests he wished made and the data already available. He also stated who would be helpful, for example, the mayor, the bank president, and the superintendent of the sugar factory in Sterling. In North Platte he had assurance that a real estate man and the superintendent of the waterworks were willing to aid. Wolff must also have appreciated the advice concerning hotels. Some were recommended, but at Ogalalla all that was reported was that "the better one is said to be on the south side of the Railroad."

When Slichter delegated, he wanted people to make use of the authority he gave them. Twice in 1905 he was irked by lieutenants whom he both liked and trusted. Reuben S. Peotter graduated from the University of Wisconsin in 1905 and probably had been one of Slichter's students. Later he did much work for Slichter along the Rio Grande. For a short period in 1905 he was left in charge at Deerfield, and on July 13 wrote Slichter on stationery of the *Wisconsin Engineer*, of which he had been business manager (the stationery proclaiming "The Engineering Department of the University of Wisconsin offers the finest technical education obtainable in the west"), suggesting that he hire at least one more man. On July 14 Slichter replied, "Perhaps I have not taken sufficient pains to make clear to you that you are authorized to incur any expense necessary to carry on the work; such as the hiring of rigs and extra men. . . . I desire to give you fullest liberty in laying out your field work."

On one occasion Wolff wanted to get rid of a young man whom Slichter had sent to him, and wired requesting his immediate removal. Slichter's reply may not have been based entirely on principle since perhaps he wished to have little to do with the case of a friend's relative. It said in part:

~ I was somewhat surprised at the receipt of this message, as I left you in full charge of the camp with full authority to hire such help at such times as seemed to be required for the best interests of the Service, and limited only by the instructions and authority covering the work which were issued last spring by the Chief Engineer. Under these circumstances I expected that you would in all these matters exercise your full authority, unless there were special reasons why I should be appealed to.

In case there is any reason why you desire to have me act instead of yourself in this case, it would be well for you to write me fully, giving the reasons that you have in mind so that I would know for just what reason I was acting, and that I might be able to discuss the matter with Mr. [X] at any time should he desire an interview. I have always made it a rule not to take any executive action in any case unless I believed myself to be fully informed in regard to all the facts in the case. You will appreciate that this is an absolute necessity.

If you have decided that Mr. [X's] services are no longer needed for any reason, please take such action as you deem best. Unless the matter is imperative, I would expect of course that every individual be given reasonable notice, so as to remove any possibility of any one believing that our action was arbitrary. I have no doubt that Mr. [X], like all men of his age, has a good many things to learn, and a good deal of patience and tact are sometimes required in handling such men. I trust you will write me more fully in regard to the matter.

Birge could be more terse. Soon after he became president, and Sellery had taken Birge's place as dean, Sellery sought Birge's advice on a particularly odoriferous problem. He got it. "George, skin your own skunk."

One interesting report dated 1906, which later was the basis of part of Water Supply Paper No. 184, had the title, "Suggestions for the

Construction of Small Pumping Plants for Irrigation." Its table of contents shows both the breadth and the minutia of the study. Among the topics are: the kind of wells, amount of water obtained, distance between wells, kind of pump and its primings, source of power, use of reservoirs, and both the initial and operating cost of the plants. The details concerning pumps probably are now of antiquarian interest. To me, still more interesting are his feasibility findings contained in the introductory remarks and in the section on "Economical Height Water May be Lifted":

~ In a large portion of the bottom lands [of the South Platte and Arkansas valleys] the distance to water is between 5 and 15 feet, making it easy to pump the water and place it upon the surface of the ground economically. The cost of pumping is controlled, primarily, by the cost of fuel, and the distance it is necessary to lift the water.

. . .

It is very unlikely that it will pay to pump water, under present conditions in the valleys of the western plains, to a total height of more than 30 feet, including the suction lift of the pump. If the pump lower the water in the wells 10 feet; and if the distance to water be 10 feet below the ground, and the discharge pipe be brought into a reservoir or flume 5 feet above the surface of the ground, the total lift will be 30 feet, if 5 feet be added to cover loss of head due to friction in suction and discharge pipe. It will probably not pay to pump water to a greater height than 30 feet at any place in the western prairies, except for the irrigation of garden truck and other high price crops.

For once he was too conservative, for now water is pumped considerably higher than thirty feet on profitable farms on the plains.

Another fine illustration of his helpfulness in applying theory to practice is a letter to Homer Hamlin, written in 1903, who had been having difficulty in driving pipes without injuring them. Slichter reminded him of the formulas for the amount of motion of the pipe as compared with the energy lost in the pipe itself and derived from these that the weight of the driving body should be as great as possible, and the velocity of the blow as slight as possible. The Archives of the University of Wisconsin contain both his handwritten draft showing how close to the final draft he could write, and the carbon of the letter showing how cursory his proofreading was, since one important line was omitted. There are many other evidences of these traits. (It is, of course, possible that in some cases the originals but not the carbons were corrected.)

Work along the South Platte was summarized in the *Fourth An-*

nual Report of the United States Reclamation Service. (I advise not using the copy in the Wisconsin State Historical Library until it is rebound; otherwise you will find your clothes covered with the dry rot of old leather.)

Although we will not review his work in Texas or Oklahoma, we quote from a letter of September 5, 1908, to Mr. Wesley Merritt about the Cimarron River Project—a letter typical of Slichter's enthusiasm, especially enthusiasm when the interests of the palate and the purse coincided:

~ *I showed Mr. McConnell the results of irrigation on the ranches of Mr. Blair F. Ruth and Jacob Behrens. On these ranches we found crops of peaches and other fruit, of the most luscious sort. The soil and climate seem well adapted to the growth of all sorts of fruit, such as apples, peaches, grapes, pears and other high-priced products, like muskmelons, and a great variety of the temperate zone plants, like sweet potatoes, etc. The peaches are unusually luscious in this valley and the crop is very certain and very early, the peaches maturing early in July. All of this land should be in a high state of cultivation, growing high-priced products for the central markets, and I so convinced Mr. McConnell.*

Slichter also made water surveys for private concerns. It could be pleasant working for a railroad in those days, for he wrote his son, Louis, on December 27, 1908, "The car is fine. I have a bedroom with brass bed and a bath room with bath tub in it. We have a fine chef and the car is stocked with the best food to be had."

Slichter did an amazing amount of writing in connection with his work for the Reclamation Service—letters of inquiry, leters of instruction, letters of promotion, miscellaneous letters, memoranda of directions, reports on individual wells, unprinted but sometimes circulated summary documents, and published reports and papers. In 1912 he estimated that his government work had resulted in "thirty thousand pages of manuscripts, notes, and data." This is probably not an understatement, for modesty is no virtue in helping an attorney qualify one as an expert witness. Considering that this mass of material was chiefly produced during university vacations, it would represent over ten pages a day—some of which presumably was data gathered by his aides.

Slichter's appointment as a consulting engineer with the United States Geological Survey ended in 1912. I find no letter of resignation, no notification of nonreappointment, nor a record of why this happened. My guess is that Slichter was bored by a repetitive job. Other interests were taking an increasing amount of attention. As described elsewhere, he cooperated with Van Velzer in writing textbooks and later published, as sole author, *Elementary Mathematical Analysis*, and became editor of a series of successful texts of which this was the first. Both as an energy-consuming and as an income-producing activity, textbook writing could take the place of fieldwork and yet allow him to be with his family. When Van Velzer resigned in 1906 Slichter became chairman of the Department, and in 1909–10 he tasted the joys of travel abroad. He was increasingly called upon to serve as a committee member. If he had not stopped his consulting work when he did, the activities entailed by the war would soon have caused him to do so.

Slichter had built his engineering practice upon his scientific investigations. He had introduced new methods into the survey of water resources. He had contributed significantly to the development of the western prairie from Texas to Manitoba and had helped southern California to come to grips with its water problem; and, if his son, Louis, is to be believed, he had been so picturesque in the process as to be offered the role of a gold panner in a movie.

City Water Supplies. It was natural that cities as well as farming communities would seek Slichter's advice. We have records of his work on the water supplies of Brooklyn, New York (already mentioned); of Rockford, Illinois; of Manitowoc and La Crosse, Wisconsin; and of Winnipeg, Canada. He was at least consulted concerning the supply of Schenectady, New York.

He did not do much work in the case of either Rockford or Manitowoc. In the first of these he was engaged by a firm including Professor Daniel Mead, an eminent engineer at the University of Wisconsin who had been a consulting engineer in Rockford, to analyze data already collected concerning the failure of some of its wells to yield as much water as expected. The most interesting feature of his report (January 1911) was the demonstration of his ability, from the curves he was furnished, to distinguish the loss of capacity due to the lowering of the water table from that due to clogging. He believed, but was not certain, that the clogging occurred in the limestone in the neighborhood of the wells and was caused by organic matter whose growth was made possible by the CO_2 introduced during pumping.

In 1911 he advised the city of Manitowoc concerning the type of study of its water supply that should be made and the most likely source of water available, as well as concerning problems of contamination.

The work at La Crosse was extensive and that at Winnipeg even more so.

The study at La Crosse arose from a petition, submitted in 1910 to the Railroad Commission of Wisconsin which also supervised public utilities, by twenty-five citizens of La Crosse who alleged that the water system of that city was unable to meet its needs, that the pressure was inadequate, and that the water coming from the Mississippi River was contaminated. Slichter was asked to report to the Commission as to the conditions at La Crosse. Trial wells were sunk under his direction, and he found a fine supply of pure underground water of not too great hardness available in the valley of the La Crosse River. He reported, on June 14, 1911, that using the Mississippi water would be risky (in the then state of purification systems) and that wells should be used.

On June 21 the Commission reached its decision, including: "The city of La Crosse is blessed as few cities in the land in that it has at its very threshhold an adequate supply of water which in quality surpasses the water of one of the most famous springs in the world [White Rock], . . . there is but one conclusion that can be reached from the investigation made and that is that the recommendations of Professor Slichter should be accepted and followed without hesitation." The order which the Commission issued was not quite as specific but, taken in context, left little doubt as to what was demanded.

Slichter was at his zestful best in studying the water supply of Winnipeg. He evidently liked the people he met and became enthusiastic over the prospects of the city. In 1912 Winnipeg had a population of about 175,000 and was supplied water from wells in and about the city. The water was high in mineral content, and if wells were to be used to meet the demands of an increasing population and growing industry, many new wells would have to be sunk. Winnipeg is on the Red River just below the mouth of the Assiniboine, which rises in Saskatchewan and flows south and then east to join the Red River. The Red River flows north, for many miles being the boundary between Minnesota and North Dakota, and is noted for its floods and its reddish brown water. Neither river was seriously considered as a source of city water. About fifty miles away is the nearest point of the Winnipeg River—a point between Kenora, where it flows out of the Lake of the Woods, and Lake Winnipeg, into which it later enters. The nearest portion of the Lake of the Woods is an almost separate branch, Shoal Lake, at a distance of nearly one hundred miles from the city. There were also some large springs, particularly Poplar and Crystal, not far from Winnipeg.

In 1907 a special commission had studied the problem and, although recognizing the advantage of drawing water from Shoal Lake, had—on account of expense—recommended the use of the Winnipeg

River. This, however, did not settle the question, and the protagonists of various sources kept the argument going. Finally, the Board of Control at Winnipeg asked the Public Utilities Commissioner of Manitoba, H. A. Robson, to look into the matter; and he employed Slichter as an expert of "highest order" but also one "entirely uninfluenced by any local interest."

Robson, at the age of nine, had come with his parents to Manitoba from England. He had been a judge of the Manitoba Court of King's Bench for two years before becoming, in 1912, the first Public Utilities Commissioner of Manitoba. He served in this capacity for three years. Later he was to serve on the Manitoba Court of Appeals, and finally, just before his death, to become Chief Justice of the province. From 1927 to 1930 he was leader of the Manitoba Liberal Party. He was, more than anyone else, not only the initiating force behind the founding of the Law School of the University of Manitoba but also a guiding hand during its youth. In the fall of 1969 the new law building of that school was named for him. It soon became evident that Robson and Slichter were kindred spirits.

Slichter's report, dated September 6, 1912, was published that autumn by the Commission. (He had visited the various sources of water during August.) He made it clear that he respected the work of the 1907 Commission. Indeed, to a large extent he based his findings on its report although, with changed conditions, he made a different recommendation. He started his report with a burst of eloquence:

~ *A perfect water supply is worth all it costs. There is no standard by means of which to measure the limit of human effort that should be expended in attaining it. The safety and permanence, and growth of the dependant civilization is too important to permit expression in terms of ordinary units. The example of Rome has been the guide to all the cities of modern cultured nations. Since that day the water supply of a city has been the most important, and the most expensive, of its public works. The abundance and purity of the water supply has determined the growth and permanence of the civic communities, and has always been a determining factor in selecting from the group of cities struggling for commercial and industrial supremacy, the favored few that should finally be awarded leadership. I trust it is unnecessary to elaborate upon the fact that a perfect water supply is the guardian and producer of wealth, or to explain in what manner a penalty, in wealth and development, must be paid as the price of a supply that falls short in any respect from what it should be or from what it might be.*

Nor did he fail to rate the various proposed sources:

> Shoal Lake—excellent;
> Winnipeg River—good;
> Crystal and Poplar Springs—fair;
> Well near Crystal Springs—poor;
> Present supply—very unsatisfactory.

The above refers to quality only.

After this stirring and forthright introduction, Slichter proceeded to an analysis of how he had reached his conclusions, an analysis which is an example of the best in exposition of a technical subject of public interest. He first forecast the future population of Winnipeg, revising upward the estimate of 1907 and assuming an almost 10% annual increase, a figure that very greatly exceeded the results.

He then took up the matter of the groundwater supply, which he pointed out did not flow at a rate sufficient to replenish what was being pumped, and he explained why the quality of the water from a sanitary point of view was impeccable. However, the mineral content was much greater than he considered tolerable for the industrial development of the city.

A sample of Slichter's ability to write lucidly follows:

~ As is well known, the surface temperature of waters, in lakes in this latitude, increases gradually during the summer months until a maximum temperature is reached, probably about mid-August. This temperature for Shoal Lake is probably not far from 65 degrees. The wind and waves keep the surface waters mixed, so that the surface temperature remains unchanged to a certain depth or layer, which lies at various depths in different lakes. When, however, a certain depth is reached (say, from 5 to 12 meters, depending upon the lake, the latitude and the time of the year), the temperature suddenly falls, often as much as 10 to 20 degrees in a single meter or two. The layer of water which possesses this rapid fall in temperature has a most important bearing upon water supply taken therefrom. Practically all the life of the lake, the minute crustacea, the diatoms and the algae, etc., exist in the warm waters above the so-called "sprunschicht." Below the sprunschicht the water is cold and practically free from living things. Even the fish exist in the upper layers, for here they find their food. Every fisherman knows this, for he seeks bass and muscalonge on the bars or in the shallow water, where the fish find their food—that is, in the upper warm water and not in the cold deep waters.

He estimated the cost of the Winnipeg River project at somewhat under four million dollars, and the Shoal Lake plan at about six million dollars. He considered this differential in costs as insignificant in light of the almost perfect water supply obtainable from Shoal Lake. Later in the report he states: "I recommend the Shoal Lake supply solely for the reason that it is the best. It is not the cheapest. I do not believe that it is necessary at this time to weigh too nicely the cost of such a project. The commission of 1907 said: 'Considered from the standpoint of the quality of the water, in its natural condition, and taking into account the use of the water for all purposes, the Shoal Lake water is unquestionably the best source of supply.' "

Slichter's comparison of the two projects follows:

~ *The 1907 commission estimated the cost of a single pipe line to the Winnipeg River, of 23,000,000 gallons per day capacity, to be $3,862,000. The introduction of electric power since that date and the changes in size of conduits, etc., that in consequence may be made, would certainly permit the construction of a single pipe line to the Winnipeg River, of a capacity of 30,000,000 gallons per day, for a sum less than that estimated in 1907. The water from the Winnipeg River must be treated and filtered for color and suspended matter, and carefully watched from the sanitary standpoint. The present population of Kenora does not threaten serious contamination. The development of water power in the Winnipeg River above the proposed intake is likely to introduce an increased industrial population, and consequently the sanitary quality of the river water may gradually become somewhat worse. If I am correctly informed of the amount of spruce and other pulp wood available near these water powers, the paper industry, among others, is one that is likely to develop. [He was right.] The waste from paper mills, even after treatment, introduces serious complications, to say the least, and the growing industrial population would require constant vigilance to maintain perfect the sanitary quality of the water supply.*

The water of Shoal Lake would require no treatment. No fear need ever be in mind that the sanitary quality of the water would be poor at any time in the future. The shores of the lake are hard rocks of the Laurentian series, entirely unfitted for agriculture, and the country thereabout must remain in its present wild state indefinitely. There need be no fear of the growth of cities or towns upon the shores of Shoal Lake. The Lake of the Woods constitutes an enormous reservoir of clear, pure and soft water, situated 300 feet above the City of Winnipeg, and within 100 miles of the city.

As previously stated, the water of Shoal Lake would never require sanitary treatment. I believe that an intake could be so located that there would be no trouble from algae. The algae are harmless from the health standpoint, but they impart a sea-weed odor and taste to water, and accordingly should be removed when present. They may grow in any artificial or natural reservoir open to sunlight.

If, for any reason, it should be determined that an intake entirely free from algae is impracticable, their removal can be effected by straining or simple mechanical filtration at the station 22.6 miles from Shoal Lake. This plan, if necessary, need be operated during a portion of the summer months only.

It is an objection to the Winnipeg River project of some importance that the water must be carefully watched from the sanitary standpoint, and filtered to secure proper purification. It is, of course, possible by proper care, to purify the Winnipeg River water, even though the industrial population at Kenora and elsewhere on the river should grow, and the present minute contamination should materially increase. If the city can command the services of suitable experts and enforce such military discipline among its employees that filtration and similar works can be operated with practically no lapse in efficiency, then it need never fear a polluted supply. In America it has been found to be very generally the case that the cities are unable to maintain that discipline and expert supervision that is absolutely essential to the operation of sanitary water-treating plants. Water supply engineers of high standing quite generally refrain from recommending such plants where a safe supply is available from other sources. "It should be fully understood and appreciated that any supply that demands filtration as an adjunct must depend for its purity on constant care and vigilance by experts thoroughly conversant with such operations, and that any carelessness or lack of vigilance will result in a temporary reduction in quality, which may, if it occurs at a critical time, result in contamination, with possible resulting sickness and death among its users." (Report on Rockford supply, 1911.) The best results, with any public work, are always secured by concentrated rather than continuous effort. A water supply which is naturally pure, and which must simply be guarded by proper construction in order to insure its constant delivery to the consumer in potable condition, is much to be desired above any supply that demands continuous vigilance as the price of safety.

The construction of the conduit line to Shoal Lake presents no unusual engineering difficulties. The lake can be reached, it seems, without rock-cutting. It appears certain that Snake Lake and Falcon River can be drained into Boggy River by a cut of about ten feet, and thus avoid about ten miles of wet excavation at the eastern terminal of the conduit. While the Shoal Lake line constitutes a major project, it is free from serious engineering difficulties. The project would be considered attractive by a large range of competent contractors.

It should be noted that, in connection with the La Crosse and the Winnipeg reports, large-scale purification by chlorination was in 1912 in an experimental state and, though more effective than earlier filtration methods, also called for disciplined vigilance.

Robson was as forthright as Slichter in his recommendations, for in introducing the report he states:

~ I recognize that one must keep within his proper province, and trust that I am not exceeding my functions when I urge that Professor Slichter's recommendations be adopted, and that the greater project be taken up immediately. The financial side of the question is of course of high importance. No one who has confidence in the stability of this city, and its future, can doubt its capacity for this project. The advantage of the undertaking should not be confined to mere corporate boundaries. A scheme might be worked out whereby the environs of present Winnipeg might, on fair terms, secure with the city the inestimable benefits of abundance of the best water. The assurance of unfailing supply is indispensable to the growth of the city. This is now the one essential to that end. Professor Slichter's words as to the future of Winnipeg are those of a widely travelled gentleman of keen observation and mature judgment. His language is not that of a mere enthusiast. Devotion to the city's present and future interests characterized every moment of the time he spent on this vital subject.

In 1913 it was decided to carry out the Shoal Lake development. Construction started in 1915. It was not finished until 1919, for a war and an inflation intervened; and the cost was nearly fifteen million dollars instead of six million. Of course the Winnipeg River plan also would have incurred higher costs than originally estimated.

I judge that there had been a good deal of in-fighting and that there was even some residual bitterness, for the Manitoba Free Press said in an editorial, on April 30, 1919: ". . . The turning of the Shoal Lake water into the domestic water mains was the signal for the stirring up of a cloud of foolish and ill-informed criticism. . . ."

The enthusiasm of the editor can be gathered from the two following quotations from the same editorial: "On the 5th of April, only a few weeks ago, the waters of Shoal Lake flowed into the mains of the City of Winnipeg. When the citizens and housewives turned on their kitchen taps, it was pure, sweet, soft, and sparkling water that splashed into their sinks." And, ". . . For so vast a project the work went forward practically without a hitch, and when the water was finally delivered to the people of Winnipeg the supply from which they drew, the water district which had been created for them, was one of the largest and most notable on the whole American continent."

On the same date the local newspaper published Slichter's paean of triumph, a letter to Robson (who by this time was Judge Robson) which is quoted in full, but with the newspaper interpolated subheads omitted:

~ My Dear Judge Robson:

I was delighted to find upon my return from Seattle a letter waiting me from you, and another came today. I had not seen any announcement of the final completion of your new supply in the public press. This surprises me, for the installation of your magnificent supply is a most important public event, and one that gives your enterprising city great honor. It is as magnificent a supply as that of the city of Glasgow, or of Los Angeles, or of New York. None of these cities possesses a supply more wholesome or sanitary, and few indeed enjoy a supply of such delightful softness. Of course all natural waters taste at first somewhat insipid compared with well waters of high mineral content. This is especially the case with soft waters containing abundant air in solution, as does your fine supply.

I judge from a clipping you enclose that some uninformed people have thoughtlessly questioned the sanitary character of your water. This is indeed a preposterous notion. No water on earth is more free from contamination, and it never will be contaminated at any future time. All natural waters are the home of living plants and animals, but only a person who would decline to drink from a mountain stream on account of the trout and cress living therein, would be foolish enough to criticize your water. There is nothing more harmful in it than salmon in Loch Lomond or the whitefish in Lake Superior.

As stated in my report, there will probably be two or three weeks in some years when the water may have a slight seaweed taste, on account of plant forms known as algae. I do not think that you will ever have to resort to mechanical filtration. Careful watching of the strainers at this time and a little care taken at the reservoir will probably be all that is required. I believe that in many seasons even this slight matter will require no attention whatsoever.

In one important respect your new supply means much more than a bountiful supply of soft and germ-free water. Unfortunately nearly all places in the central continental area of North America that rely upon

*well water suffer to a material extent from goitre and thyroid trouble.
Our recent examination of young men for the army showed this to be
quite a serious matter in parts of the States in the central area. Now
Winnipeg has not been altogether free from this trouble in the past.
Now, however, you will be able gradually to note a change in this situation, and a progressive diminution of thyroid trouble will be in evidence
as your glorious supply of natural water has time to have its effect.*

*I shall certainly do all I can to spread the news of the completion
of your magnificent supply of water. It should be a source of great pride
and profit to all of you, and I hope the city is not too modest in letting
the world know. Los Angeles and New York seemed to show no delicacy
in letting the world know of their water supply accomplishments. That
great pure supply of yours, from those old Laurentian granite rocks
around that beautiful lake, is something worth boasting about—you
need not fear comparison with any supply of any place on earth.*

> *Very sincerely yours*
> *Chas. S. Slichter*

It is not surprising that Slichter did not know the relation of iodine
to goitre. Although some doctors had used iodine in treating goitre as
far back as the middle of the nineteenth century, it was not until 1917
that the first large scale experiment in preventing human goitre by the
use of iodine in the public schools of Akron, Ohio, was reported. It
could hardly be expected that an engineer would be completely au
courant of medical development, although it may have been slightly
naive to assume that the purity of a city's water supply would be
adequate insurance against the disease. The use of iodized salt soon
became common; and if the incidence of goitre in Winnipeg actually
decreased, undoubtedly it should be attributed to a substance, other
than water, that also pours when it rains.

Shoal Lake is still the source of water for Winnipeg and the city
proudly boasts of the fact. However, the "glorious supply" of water from
Shoal Lake is now both chlorinated and fluoridated.

Trials Without Tribulations. Slichter was frequently consulted on
legal questions concerning water rights. For instance, he furnished
considerable information to Isaiah Bowman (at that time Professor at
Yale and later president at Johns Hopkins) concerning the validity, or
rather lack of validity, of the claims made by some Long Island farmers that the Brooklyn water supply system was robbing their wells.

As stated in Chapter 7, he tried not to testify as an expert witness
but at times was unable to avoid doing so. Chapter 8 gives his account

to Town and Gown, as described by a fellow member, of experiences as a witness in Colorado. This was in connection with a Kansas-Colorado controversy.

On another occasion he was a witness in Seattle. His letter to his son, Sumner, written February 9, 1922, follows. It is a good conclusion to this chapter, for Slichter was a man who got fun out of his work.

~ *I finally got my testimony in and left this morning. The case will run two weeks more. I can not tell the outcome. I had a fine time on the stand and had a lot of fun with the lawyers because they were so ignorant of underground waters. The judge is a fine able fellow and enjoys a joke very much, and I had a good many chances to cheer him up. The jury had not heard a joke for three weeks. The lawyer on cross examination asked me the ultimate origin of all waters—rivers, lakes, and springs. He wanted me to say "the rain." I said "the seas." He then said, "Isn't the ultimate source the rainfall?" I then quoted: "God said, let the waters be gathered together in one place, and the gathering together of the waters called He the sea; and the morning and evening were the third day." The jury roared. He was then foolish enough to read from one of my publications where I said that the source of all surface and subterrane waters was the rainfall. He asked me if that were not true. I said "Who said that." He said, very decisively, "Slichter." I then said, "I have quoted what God said. You have quoted what Slichter said. There is a conflict of authority." I waited a minute, until they had laughed several minutes, and then said, "Slichter is right." I had no trouble with the jury after that.*

Chapter 5 ❧ *University Citizen*

A great institution has its own flavor which some savor and others find repugnant. At the University of Wisconsin this flavor has been derived in large part from faculty citizens who cared about the University as a whole and who were willing to take the lead in committees, in faculty discussions both formal and informal, in collective enterprises for the good of the University community, and in representing the University before the public. Some of these were of the steady, devoted, wheelhorse type, but to a greater extent they were men of distinct individuality—colorful characters—with many idiosyncrasies. In Slichter's day they were such men as Van Hise, who, after warning students of the dangers of Lake Mendota, would skillfully handle a canoe upon it in a storm; Birge, with his "pirate's eye"; Russell, an indigenous bear with a steel-trap mind; Hohlfeld, inheritor of the European scholarly, courtly, but pugnacious traditions; Howard L. Smith, whose rapier wit reduced formidable persons to impotent rage; McGilvary, the agnostic son of a minister, who, however, clung to the Calvinistic ethics; Bradley, the acme of the competent outdoor man; Commons and Ross, let go from other institutions for noncooperation; a goodly number of others; and Slichter, "not least, but honored of them all."

This chapter is devoted to Slichter's activities as a faculty citizen. Although it is forced to be as heterogeneous as his various interests, it will indicate an amazing energy expressing itself in many ways for the good of the University.

Committees

There were a number of committees on which Slichter served in an ex-officio capacity as dean of the Graduate School. These are described in the chapter on his deanship or not at all. There were plenty of others.

There is no point in discussing extensively his services on a seemingly endless series of committees. I will, however, discuss in some detail his work on four committees which dealt with student affairs and welfare—namely, those on athletics, the *Daily Cardinal*, the dormitories, and the Union—and more briefly his activities on some of the others of which he was a member, in order to give some idea of their variety.

Athletics. Although Wisconsin had played its first intercollegiate game—baseball against Beloit—in 1873, Slichter's service on the committee in charge of athletics almost spanned the chaotic growth of intercollegiate athletics at the University. It was inevitable that a popular teacher who liked athletics and enjoyed being with students would become involved.

The general development of athletics at the University of Wisconsin is described in the history of the University written by Curti and Carstensen. Professionalsim seemed to have been even more rampant and more deplored, especially by opponents, in the late nineteenth century than it is today. Rowdyism in athletics was at least as bad then as it is now. The degree of financial irresponsibility was almost unbelievable.

We shall start with the end of the story: More than a year after he had gone off the Athletic Council, and while certain reforms and the faculty control of athletics were in a trial period, Magnus Swenson and Lucien S. Hanks wrote, for the executive committee of the regents, to Slichter and presumably to a number of others of "the older and stronger men of the faculty," asking their opinions on the "effect of intercollegiate athletics on the University as a whole and on the student body." The major portion of Slichter's reply, dated April 12, 1907, follows:

~ I suppose there is little difference of opinion concerning intercollegiate athletics as formerly run in this University; the effect was undoubtedly bad. However, as a result of the reform measures taken by the Faculty of the University and the Conference, it is believed that the evils connected with intercollegiate athletics are in large part removed. It is equally true, in my judgment, that intercollegiate athletics supply an actual need in the University; the question has always been between the benefits and the evils. Now that full control of athletics has been given to the Faculty by the Regents, I cannot doubt that such control will result in the development of intercollegiate athletics of a type that will be entirely satisfactory to the Faculty as a whole.

In any case, and independent of any views a person may have concerning the probable success of the present reform measures, I believe it to be our duty—both of the Faculty and of the Regents—to cordially support the present program until the new regime can be thoroughly tested out. If we do not give cordial support to the athletics while they are under trial, as at present, we will be met with the claim—if athletics turn out to be unsatisfactory—that those opposed to athletics have contributed to the failure by their unwillingness to give cordial support to those that were attempting to carry the burden of the reform. One of the regrettable things of the present situation seems to be that some continue to find fault in a petty way with the conduct of athletics; the more carefully the reform ideas of the Faculty are carried out, the more they deem it necessary to criticise and find fault. Just the contrary spirit should prevail if we expect the test now being made to be of any value.

It is well known that Professor Turner, in consultation with the President, has been exceedingly active and has done great work in accomplishing the reform of intercollegiate athletics as proposed by the Conference and the Faculty. My loyalty to him, and my recognition of his great services in this respect, would alone require me to cordially support the present athletic regime, no matter what my personal views might be. We cannot expect men to go to the trouble that the President and Professor Turner have gone to without supporting them for the period for which they have virtually asked such support.

If my own personal views have any value, I must say that I believe in intercollegiate athletics, with the exception of intercollegiate baseball. I believe that they can be run so as to eliminate the unwholesome features and be of great use to the students and University. I am entirely optimistic concerning the outcome of the present trial.

I trust that the Regents will give most cordial financial and other support to intercollegiate athletics, and will sustain the Faculty most willingly in the present adopted program. Any attempted change in the present regime before January 1, 1908, will lead to trouble, misunderstanding, and unfortunate complications in the University. It is absolutely necessary to keep good faith with the students in the matter.

(At the time this letter was written, baseball was the only sport on a sound professional basis.)

The prelude to this letter had been an intimate and continuous connection with athletics for a long period of time. In 1889, three years after Slichter arrived at the University, the faculty established a Committee on Athletics whose chief duties seemed to have been connected with schedules and eligibility. The first committee consisted of Professor Edward T. Owen, Professor of French and Secretary of the Faculty, as chairman; Lieutenant James Cole, of the Military Science Department; and Slichter. In 1894, for an unrecorded reason, this committee was changed to a council, with perhaps somewhat greater powers but still far short of "faculty control." The first council members (the faculty minutes do not say who was chairman) were Dr. Elsom, Lieutenant Cheynowith, Professor Barnes of Botany, Slichter, and Van Hise. Slichter was made secretary at the first meeting and held that position for about two years. By the fall of 1900 he was chairman. The first University catalog that listed the standing committees of the faculty, 1903–4, included the Athletic Council with Slichter as chairman, five faculty members, "and four students chosen by the athletic board"—which must not be confused with the board that now bears that name. At various times Slichter was called "Supervisor of Athletics," probably an ex-officio title borne by the chairman of the Athletic Council.

In 1905–6 Slichter was replaced by Professor Thomas S. Adams of Economics, and no students are mentioned. Slichter had resigned from the council. Perhaps in light of constant faculty attacks on athletics, especially football, it was deemed wise to have the leadership of one less openly favorable to athletics than Slichter had been.

At such a time there was naturally some good-natured, but not entirely biteless, banter between two of the leading wits of the faculty— Slichter and Howard L. Smith of the Law School, one of the sharpest critics of the athletic establishment. On April 22, 1905, Slichter wrote to Smith:

~ *I enclose herewith a communication from one [O. P.], who alleges that he is an amateur, middle weight, champion wrestler of the state of Wisconsin.*

This man seems to desire to take up the course of law, and desires to come to some agreement for the payment of his expenses. You will kindly note what he says about the interest that he has taken in physical culture, and that he claims to be a specialist in this line. I know of no one that I would rather apply to in this emergency than you.

I trust that you will take this matter up with [Mr. P.] and do anything you can to add to the glory of the Wisconsin Law School in things physical as well as mental.

The reply was dated April 28:

~ *My dear Slichter*

I regret that the funds available for this purpose for the current year are already bespoken.

If it be not lèse majesté I would suggest that [Mr. P.] apply to

*Michigan. I understand that the higher learning is more liberally sup-
ported there than here—at least since the chief apostle of the higher
learning was dropped from the athletic board here.*

There was as constant a reorganization of the student role in the
athletic picture as of that of the faculty. In 1901, for instance, there
was a "University of Wisconsin Athletic Association" whose membership consisted of "all students, former students, and members of the
faculty of the University of Wisconsin." The Association had a governing board consisting of its president, vice-president, secretary, three
members of the faculty, one alumnus, one regent, and ten undergraduates. Both intercollegiate and intramural athletics were the concern of the Association. One feature that now would seem strange was
the rule, "The crew, the baseball team, the football team and the track
team and all substitutes to the same shall be selected by the captains
of the respective teams." Naturally there were detailed rules for the
selection of the captains. Some inkling of previous arrangements was
given in Article X, "This constitution shall go into effect immediately
after its ratification by the Football, Tennis, and Baseball Associations
and the Boat Club or by any three (3) of them."

Whoever was the chief representative of the faculty, in dealing with
intercollegiate athletics, had to advise students as to what activities
would make them ineligible because of professionalism, and had to
play a leading role in ferreting out professionalism within opposing
teams.

A few examples follow: In 1904 a teacher from near Albany, Wisconsin, who expected to return the next fall to the University and
"try for the football team," wanted to play on the school team with
his students against Brooklyn under an agreement by which the expenses would be paid by the losers. He asked if this would jeopardize
his eligibility. Slichter replied:

~ *The conditions that you state in your letter governing the proposed
football contest are sufficient to render all of those who participate in
the game liable to the charge of professionalism. The $25.00 you name
is of the nature of a purse which is played for by the contesting teams,
and since the payment is to depend upon the winning or losing of the
game, it would clearly violate both the letter and the spirit of the amateur rules.*

However, the four boys who wanted to play for tennis clubs during
the summer were reassured by the following letter of June 17, 1905.
"Your note in regard to Tennis is at hand. In reply will say that tennis

and golf and other gentleman's games have not been considered as coming under the direct regulation of the athletic rules."

In the same year Slichter wrote to a judge in northern New York concerning a Michigan student who had "no visible means of support," and received a detailed record clearly indicating that the player was a professional and the likelihood that Yost and others at Michigan knew the facts. There is a hint of smugness in Slichter's thanks:

~ *I have your letter of Sept. 28th for which I thank you very much.*

I am greatly amazed at the situation that your letter depicts. I do not believe that we are justified in permitting [Mr. X] to play any further in case he is returning to Michigan.

This should be compared with his letter to Edward S. Jordan, who was collecting material on athletics in some midwest universities for an article to be published in *Collier's Weekly*. After rather evasive replies to a number of questions, Slichter concludes:

~ *I trust that you will have a very good time in collecting the information that you desire. I greatly regret that I have so little to offer that would be of any assistance to you. I have no opinion that conditions are any different at the other universities than they are at Wisconsin, unless it may happen to be that better judgment has been exercised in selecting players at some of the other universities than at Wisconsin. Results might make this last seem probable.*

An interesting sidelight showing some willingness to look the other way was given also in this letter:

~ . . . *The record of the Minnesota team and the character of their men is the best criterion on which to judge of [Mr. X], if we grant that it is fair to judge a faculty man by the character of the men they get on a team. I do not believe, however, that this is quite a fair way of judging a man, as the faculty members of an athletic committee probably would be the last ones to receive any hint that members of the team were ringers.*

These quotations are particularly interesting for, probably unbeknownst to Jordan, Slichter had written on January 19, 1905, to Richard Lloyd Jones, Associate Editor of *Collier's Weekly* and a graduate of the University, about the possibility of Jordan's writing on athletics. After praising Jordan, he continued:

~ . . . *As Chairman of the Athletic Council at the University of Wisconsin, I know something of the difficulties connected with getting at*

the facts. As soon as a member of the faculty is appointed on the athletic committee, it seems to become everybody's business to keep him in
as great ignorance as possible concerning the real situation, both at his
own university and at others. After the athletic season is closed, a good
many things come to light that prove to such a person the great extent
to which he has been kept in ignorance.

I have a strong conviction that publicity should be given to the situation in intercollegiate athletics in the west as well as in the east, and I
would be glad to see Collier's Weekly take the lead in such a movement.

There is some evidence that members of the Athletic Council themselves participated in "everybody's business." For instance, when in
1901 a football player was accused of professionalism, a *Cardinal* re-
porter interviewed Slichter who said: "I don't know anything about
the [X] affair, except what I saw in this morning's paper. Minnesota
has not formerly protested him. You can say about [Z] that Wisconsin
is not stirring up an agitation against him, but that Minnesota authorities are looking up the matter themselves."

In November and December, 1905, Jordan published four articles in
Collier's Weekly. The situations at Chicago, Northwestern, Illinois,
Michigan, Wisconsin, and Minnesota were discussed. One comment
read, "At Wisconsin nominal power rested with Professor C. S. Slichter, Supervisor of Athletics. This man believed in absolute athletic
decency, but lacked the nerve to use his absolute power." Except for
President Van Hise, few persons at any of the institutions were treated
in as complimentary a manner.

Of course it was chiefly the better athletes who would be offered
opportunities to play for money, and they were the most easily de-
tected. Assumed names were often ineffective protection and added
the real turpitude of deceit to the morally innocent practice of receiving money. Occasionally the ineligibility of such athletes was removed after a sufficient period of penitential poverty; and even long
after college careers were over, athletes sought reinstatement as a sign
that any fault was fully atoned.

Financial crises seemed to have recurred with great frequency. No
description of these is more vivid than that given by President Adams
to the regents on April 17, 1900.

~ *I see no reason for changing anything I have said in regard to athletic matters. The reign of politics among the students is unabated, and
the Athletic Board recently chosen appears to be in hearty accord with
the business manager, whose accounts were found by Professor Slichter
in so chaotic a condition that he declared they could not be audited.*

They also baffled the skill of Mr. Burd in the Treasurer's office; and Professor Van Hise, after working over them for some weeks, was obliged to leave them unsettled. No public statement of the result has yet been made. It is evident that very large numbers of those who heretofore have subscribed liberally for the boat crew and for other athletic matters will do absolutely nothing until a change of one kind or another in the directorship takes place.

It was not until 1907 that control of the finances of athletics was granted by the Regents to the faculty, and delegated by them to the Athletic Council.

Not unrelated to finances was the perennial searching, in the later years of Slichter's connection with athletics, for coaches—especially football coaches. This was in the hands of the Association and its board rather than the Athletic Council. The alumni were often critical of either the choices made by the students or the length of time it took to reach a decision.

When, in 1904, there was a vacancy in the position of Graduate Manager of Athletics, Evan A. Evans of Baraboo, a graduate of 1897 on his way to eminence in the legal profession, wrote to Slichter recommending a classmate for the position. He expressed himself as many a graduate of many an institution has done before and since.

~ *Like many other alumni, I have attended the athletic contests in the last few years and have occasion to regret the failure to successfully compete with the colleges that formerly were not our equal. It seems to me that we must be especially careful in the selection of all coaches and managers in order that the best results might be obtained. I fully appreciate the great benefit of University athletics, is not to win the various contests but at the same time, it seems for the wages we pay, that we ought to be able to secure the services of men whose influence over all students would be for the good and at the same time secure efficient services in the line of a coach.*

Slichter answered:

~ . . . *I was very glad to hear from you, as I believe it is very important that the alumni should take an active part in the athletics of the University during the present important crisis in our affairs. We are very much interested in securing a strong and efficient man for this position, as well as for the position of coach of the football team. In order to accomplish the best results it will be necessary for the students to have the advice and assistance of all loyal graduates.*

The loyalty of Evans persisted, as did his willingness to criticize. The foregoing purring reply was just practice for the correspondence with Evans more than two decades later when he was a judge of the United States Circuit Court and a trustee of the Wisconsin Alumni Research Foundation.

Slichter would occasionally stir up alumni when he believed that the Athletic Board (he was on the Council) was procrastinating. On January 18, 1905, he wrote to Lewis L. Alsted, a Milwaukee alumnus.

~ I have a feeling, based upon the situation as it is known to me, that very little is being accomplished in the matter of the selection of a coach. There does not seem to exist an appreciation of the importance attaching to this matter, or any knowledge of the rather delicate questions involved. . . .

While I have no responsibility in the matter, and can not alone do very much to impress upon the local committee the importance of prompt and thorough work, yet I do feel that the alumni, resident in Milwaukee and Chicago, should know that things are not being pushed, so that they may do whatever they can to stimulate matters.

It may be, of course, that I am unduly exercised about the matter and that everything is going along all right and the best things will result. My conviction, however, is very strong that the situation is quite contrary.

Alsted was already incensed but judged that all that could be done was to let matters get worse until there was an explosion and a reform —however, not the kind of reform advocated by the antiathletic element in the faculty. He also foresaw some of the financial problems that losing teams then and now entail. Hindsight could well have been the basis of his foresight.

A task of the "supervisor of athletics" was answering, often for administrators, questions which came from outside the University. On May 9, 1905, Slichter wrote, for President Van Hise, to J. H. Farley, Professor of Philosophy at Lawrence University, answering a list of some ten questions. The answers to questions 8 and 10 follow:

~ 8. We have no special rules regarding betting. The statutes of the state are ample in such cases, and notices have been posted by the President of the university on the athletic field before athletic contests. . . .

10. We have no definition of legitimate work which a student may offer, if he is working his way through school.

On November 28, 1904, he wrote to Professor D. C. Hall of the University of Oklahoma:

~ . . . In reply I would say that the football coach is hired by the Athletic Association of the University of Wisconsin. He is paid a salary of from $2000 to $3000 a year, or for the football season, and is paid by the above Association. The University pays no coaching salary to coaches, and assumes no financial liability on account of University athletics.

On October 25, 1904, at the request of Dean Birge, he replied to A. H. Sproule, Assistant Principal of the Elgin High School:

~ Rules governing personal conduct of men of the team are entirely in the hands of the coaches and trainer and I believe are not put in form of a written code. The rules require abstinence from tobacco and intoxicating drinks, and the close observance of regular hours for meals and sleep. The men are boarded at a special training table and pastry and other nonnutritious foods are cut out of the bill of fare, and meats and other foods supposedly of high athletic value are substituted for them. It is my impression that the severe rules to which the men are obliged to subject themselves, in order to participate in athletics, would be of considerable value, if it were not for the fact that the force of these rules suddenly terminates at the close of the athletic season. This has a very bad effect upon the physique and I believe upon the [character] of the man. The intense athletic work of the football season is followed by a period in which the exercise the man takes is entirely a matter of voluntary desire. The result is that many of them do not take sufficient exercise. I find that the training is not only bad from this side, but in some cases it is even worse, as I believe the [end of] training is apt to take the form of a prolonged debauch.

Another chore of the Council was exempting athletes from gymnasium and military drill, and even—until it begged off from the responsibility—excusing women students from courses in hygiene.

After Slichter was off the Council his interest in athletics remained. As witness—on December 20, 1906, he wrote to C. P. Hutchins, Professor of Physical Training:

~ I understand that the Athletic Council has approved a schedule of baseball for the coming spring. I notified a member of your committee that if such a proposition was made I would oppose it before the University Faculty, and I asked him to make known to you my intention so that you could so have matters in mind that the question could come to the attention of the faculty without prejudice.

Last year we had more baseball than ever before and the students had a great deal of enjoyment out of it. This was largely on account of the absence of a baseball team playing intercollegiate games. We have thus proved *that intercollegiate baseball is absolutely unnecessary for athletics in the University of Wisconsin.*

Slichter was evidently mollified, for at the January 7, 1907, meeting of the faculty Dr. Hutchins announced the approval by the Athletic Council of the baseball schedule and, although the question remained a live one, the only record of Slichter's participation in the meeting was in regard to entrance requirements.

More representative of his attitude towards athletics was what he said at a convocation on October 30, 1903. We quote the *Daily Cardinal* of that date:

~ *Prof. C. S. Slichter, faculty supervisor of athletics, in his remarks on the same subject, maintained that the great interest in athletics today was not merely interest for the sport itself. The abnormal growth of university athletics is not accounted for so readily. There must be some underlying cause and that cause is the natural qualities of any community, great or small. All nations of antiquity had festival days, which were merely occasions for an inner spirit—which demanded expression. In like manner there is a spirit at the university—which demands that the institution may repress [sic] itself as a unit. The great interest in athletics today is demand on the part of the students, as a body, to show that they are part of a great unit, the university. We appear at a football contest not as individuals, but as members of a great university.*

We must conclude that, by and large, for nearly two decades Slichter lived in a state of symbiosis with the organized confusion of intercollegiate athletics.

The Daily Cardinal. It is not the purpose of this section to trace the vicissitudes of the *Daily Cardinal* but rather to describe Slichter's connection with it.

The *Cardinal* started publication in the spring of 1892 as the first daily student newspaper at the University. It had been preceded by various periodicals, partly of literary and partly of reporting natures. The relation of the *Cardinal* to the faculty and to the Regents has seldom remained constant for any long period. The faculty has attempted to have some influence on the *Cardinal*'s content, even at times having an official "censor of student publications." During much of the life of the *Cardinal*, the Regents have either granted a subvention to it, paid it for providing official notices, or sent subscriptions to the high schools of the state. At times it has been designated as the official paper of the University. The financial structure has also varied.

For a number of years before 1914 it was owned by a stock company with the stock owned largely by faculty members, alumni, and Madison businessmen. The company had a governing board chosen from the stockholders, and it was as a member and officer of this board that Slichter served.

He does not seem to have spent much energy upon the publication policy of the paper but, as in the case of the Athletic Association, was active in trying to unravel the constantly recurring financial snarls that various editors and business managers got into.

We know that he was active in the *Cardinal*'s affairs as early as 1897, for there is an old ledger with a note at the end of the 1896–97 entries: "Accounts in accordance with report of Mr. Slichter. J. B. Sanborn." John B. Sanborn at that time was a young lawyer—destined for a distinguished career as a member of the Madison bar. His father, Arthur L. Sanborn, was a lawyer, and from 1905 until 1920 was United States District Judge in Wisconsin. (I have one fond memory of John Sanborn for, at one of the first meetings of the Madison Literary Club which my wife and I attended, he pronounced Oklahoma the way I and my geography teacher did, namely: the "Ok" to rhyme with "knock" and the "homa" with the Madison pronunciation of "Homer," instead of the way that my wife and most Madisonians do.) We also know that Slichter's connection with the *Cardinal* continued at least as late as 1909, for in February of that year President Van Hise had a hot note from William S. Kies, who evidently had underwritten some of the debts of the *Cardinal* and, according to him, was rudely sued for money by the Cardinal Association. He said that his communications had been ignored. He declared that he was "a loyal and enthusiastic" alumnus but that thoughtfulness on the part of the president and "the other professors who are associated with you on the *Cardinal*" might make him "even more loyal and enthusiastic." The President asked Slichter to report on the matter. We do not know the outcome of this episode, but Kies' loyalty and enthusiasm, like that of Judge Evans, remained undiminished—for, as described elsewhere, it was to Kies and a few others that Slichter turned in 1917 to get money for Max Mason's submarine detection work and later to help get the Wisconsin Alumni Research Foundation started.

For the years between 1897 and 1909 we have several bits of correspondence between Slichter and the editors or the business officers of the paper. For instance, one year he told a student that he had been elected manager of the *Cardinal* and informed him that his compensation would depend on his success in reforming the system of distributing the paper, pledging the help of the Board of Directors in doing so.

A year later he wrote an almost brusque letter reminding the young man of his obligation to wind up his accounts. Another letter responded with a "No" to a request from a *Cardinal* staff member for more pay.

No one has accused Slichter of being a patient man—but he was able to live for many years with situations that stirred his ire. Patience and persistence are different qualities.

Dormitories. Slichter's work on the dormitories and on the Union came later than his involvement either in athletics or with the *Cardinal*, and much of it was done (but not in an ex-officio capacity) while he was Dean of the Graduate School.

Although during the first few years of the University parts of North Hall had been used for a men's dormitory, for many years before Van Hise became president there had been practically no dormitory facilities for men and only Chadbourne Hall for women. Van Hise strove to increase such facilities and succeeded in getting funds with which to build Barnard Hall as a dormitory for women and Lathrop Hall as a women's social and athletic building. Although no dormitories for men were built during his administration, studies were made and campaigns to secure legislative support were initiated.

Some felt that many of the suggestions were too posh. Slichter was among these and believed that the greatest need was for low-cost housing.

On December 2, 1914, the Regents voted, "That the president of the university appoint a committee of three from the faculty to cooperate with the business manager and architect to present alternative plans for the men's dormitory of a cheaper type of construction, to be presented at the next meeting of the board: the question of dining room facilities to be included in their studies." Van Hise had evidently given the matter some thought, for he immediately announced his appointments: Professor John G. D. Mack, who the next year was appointed State Engineer; Dean Turneaure, of the College of Engineering; and Slichter, as chairman. He asked that they work swiftly, suggesting that, if possible, they report to the Regents in two weeks. We have not found a report and it may have been oral; but on January 20, 1915, the Regents authorized continued study by the Constructional Development Committee of a building that would cost $450 for each student housed.

This was not Slichter's idea. In the March 1915 issue of the *Wisconsin Alumnus* he published an article entitled, "Shall We Reduce the Cost of Living at the University?" He did not refer to the Regents or to the special committee which he had chaired, but he urged a plan whereby he believed that the dormitories could be built at $300 per student, and with such dormitories a student could receive "wholesome but simple board, clean, sanitary rooms—one room to each student" for $150 a year, or perhaps as low as $126. He concluded:

~ A plan like the one proposed above is quite essential if bright, ambitious boys, whose parents possess small means, are to receive the advantages of a university training. A farmer who has a boy 17 or 18 years of age is probably in middle life—say from 40 to 50 years of age. Such a man very likely still owes something on the farm, and probably has need for more funds for better stock and better buildings. One hundred fifty dollars per year is all that he can reasonably afford to contribute to the boy's higher education. I would like to see it brought about that a clever young man, supplementing aid from his father by work during the summer vacation, could live comfortably at the University, with the privilege of devoting all of his time while there to the pursuit of his studies. He would make much better progress toward realizing his ambitions than he can at present, when too often he must spend so many hours in work for self-support that his university time is cut down to a minimum and his success is endangered.

If Wisconsin solves this problem in only a half-hearted way, some other State University will take it up and do the work much better than we have done it. This kind of competition is a good thing for us; it should make us sufficiently cautious so that we shall not work out a solution that will be easily beaten by another State University. Very excellent results have been reached by the numerous colleges and universities in Ohio and Indiana. In these states there are so many competing colleges that each college must handle the problem of the cost of living very thoroughly and about as well as any of the other colleges in the competing circle. Many of these colleges, such as Miami, Oberlin, Ohio University, Western Reserve, etc., are high grade institutions with excellent reputations. Yet a number of these still advertise board and room for $135 to $150 per student. I wish that we at Wisconsin felt their competition more keenly. It would wake us up from our easy going ways and would stimulate us to carry on an actual program of research on the value of the dollar and on the need of thoroughgoing business management.

Eleven years later, in 1926, when during a period of inflation new men's dormitories opened, room and board in fact cost about $372 a year.

In 1924 the residue of the estate of J. Stephens Tripp became available to the University. Tripp, who had come to Wisconsin from New York State as a young man, had been for a long period a leading citizen of Sauk City and Prairie du Sac—first as a lawyer and then as a banker, with periods as Assemblyman and Postmaster. He died in 1915 at the age of eighty-seven. His will provided for a number of specific gifts and the support of two sisters and a brother, the residue to go, at the death of the last of these three, to the University of Wisconsin. The University ultimately received over a half-million dollars from the estate. The Regents set aside $300,000 of this, to be repaid if possible (repaid in 1968), for financing the first two men's dormitories—Tripp and Adams Halls. These were erected on the shore of Lake Mendota and many (including me) regretted that the site chosen was that already occupied by a magnificent group of weeping willows.

Harold C. Bradley, chairman of the "Committee on Undergraduate Social Needs," and Slichter were leaders in getting the dormitories started. Bradley was Professor of Physiological Chemistry, and after retirement was a leader of the conservation movement in California. Another member of the committee was Otto L. Kowalke of the School of Engineering. Kowalke credited Slichter with the leading role not only in planning the buildings but also with two special features: the housefellow system and the naming of the separate units (houses) within the halls after prominent alumni, and securing portraits of these persons to hang in the lounges of the various units. In an interview with Allmendinger on January 11, 1966, Kowalke described Slichter's leadership: "How Slichter ran the Committee: He would never reject proposals of other members outright. He would put them up for critical discussion, and lead that discussion. When he made proposals, they were completely researched, planned, and forcefully presented. 'I just followed his leadership. He was the Boss.' Slichter carried weight. He had the ability to suggest things and make them seem a part of the group's ideas and carry them this way."

Bradley's role was much more important than the above would indicate, and he should get credit with Slichter for the housefellow plan. However, Slichter was the chief protagonist of the plan. Porter Butts, for many years Director of the Wisconsin Union, reports that, when he became a member of the Residence Halls Committee in 1926, he learned that it was Slichter, "enamoured of the Oxford don system as he was, who was largely responsible for Wisconsin's first halls for men being not 'dormitories' but small house units, each with a common room, or den, each with a 'fellow' in residence. I can still hear him talking, in committee and out, when questions were raised about the expense of the small units and the living room suites for fellows: 'These halls aren't just for sleeping and eating. They will be an experience in living and learning together—and that's an important part of a young man's education.' "

The naming of buildings and the securing of portraits were carried to success with great enthusiasm by Slichter. He consulted many persons and got enough contradictory advice to give him great freedom to reach his own conclusions.

Judge Evans, as always, had definite, and in this case prophetic, opinions. After suggesting the names of Tripp and of Frankenburger, former Professor of Rhetoric and Oratory, he stated: "I am opposed to its being called Van Hise Hall for the reason that Van Hise is so well known and has been so fully honored that it is unnecessary to attach his name to a dormitory. Likewise, there will come a time, no doubt, in the history of the University when a building of larger importance may well be named after him. Let us honor the more modest and unknown professors whose work, however, in the opinion of many, and with many, would be greater than that of the individual who occupies the presidency." He had the same objection to naming dormitories for either Vilas or La Follette. The final decision was to name the two halls after Tripp and President Adams.

The committee followed Slichter's recommendation (which gives an inkling of Slichter's taste in heroes) on naming the houses, and on May 18, 1927, the Regents took the following action:

~ That the sixteen houses in Adams and Tripp Halls be named after the following alumni as recommended by the Dormitory Committee:

1858—William F. Vilas, A soldier of the Civil War, United States Senator, and for many years a Regent of the University of Wisconsin and a substantial benefactor.

1859—Samuel Fallows, A soldier of the Civil War, a Bishop of the Reformed Episcopal church and a most loyal alumnus.

1859—Alexander C. Botkin, Editor of the Chicago Times, Editor of the Milwaukee Sentinel and codifier of the Criminal Laws of the United States.

1864—James L. High, Soldier in the Civil War; very prominent at the Chicago Bar, and author of several important legal treatises.

1864—John C. Spooner, Prominent attorney and United States Senator. A soldier of the Civil War.

1869—David B. Frankenburger, A professor in the University and an active and loyal alumnus.

1870—Stephen S. Gregory, A distinguished attorney, an ardent and devoted alumnus, President of the Bar of Chicago, President of the

Bar Association of Illinois, and President of the American Bar Association.

1873—James W. Bashford, a bishop of the Methodist Church, for many years a resident of China, and President of Ohio Wesleyan University.

1873—George H. Noyes, Circuit Judge, long a regent of the University, and an ardent friend and supporter of the institution.

1878—Robert G. Siebecker, Attorney and Justice of the Supreme Court of Wisconsin.

1879—Robert M. La Follette, Attorney and United States Senator.

1879—Charles R. Van Hise, Geologist and President of the University.

1880—Henry B. Faville, A distinguished physician and a loyal friend of the University.

1880—Henry L. Richardson, A clergyman of the Congregational Church. A hero of the Iroquois Fire of December 30, 1903. He repeatedly entered the burning building and brought out a number of little children. His body was found in the ruins with two little children in his arms.

1884—Albert J. Ochsner, A distinguished surgeon and active alumnus.

1890—Warren D. Tarrant, An able attorney and Judge of the Circuit Court of Milwaukee, an earnest friend of the University.

Soon a new refectory was named for Van Hise; and the house was renamed—probably at Slichter's suggestion—for John B. Winslow (1851–1920); undergraduate, Racine College, 1871; Law, University of Wisconsin, 1875; Justice of the Supreme Court of Wisconsin, 1891–1920 (Chief Justice, 1907–20). The name of the refectory was changed to the Frank O. Holt Hall when the present language and administration building was named for Van Hise.

Slichter's activity in getting portraits is exemplified by the letter he wrote to Van Hise's daughter Janet, at one time a next-door neighbor of his on Frances Street.

~ My dear Janet:

You probably know that I have taken a great deal of interest in having the various Houses which constitute the University men's dormitory system named after prominent alumni who in their day served their generation in a distinguished way.

I have been able to secure oil portraits of the persons after whom these houses have been named for the club room in each house in a large number of cases, and I desire to put through to completion during the current year the acquisition of portraits for all of the Houses.

One of the Houses, as you know, is named Van Hise House after your father, and I am writing this letter to suggest that you take up with Alice and others the possibility of giving the University an oil portrait of your father for Van Hise House. We have found that these portraits are very much cherished by the boys living in the Houses, and I know that the influence of the strong faces looking down upon the boys in their club room is profound.

As each House holds only thirty or thirty-two students, the club rooms are ordinary sized rooms with rather low ceilings, so that a portrait not larger than about thirty by forty inches, including the frame, is the best size for this purpose.

I would be very glad indeed to talk this matter over with you after you have had an opportunity to think it over.

He was successful in getting the portraits of a large majority of those for whom houses were named.

Conditions were drastically altered by the establishment of the Experimental College under Alexander Meiklejohn. Certain of the houses in Adams Hall were assigned to the Experimental College. It was planned that during the next year, 1928–29, all of Adams Hall would be used for this purpose. On March 5, 1928, Bradley sent to Slichter a draft of a letter which he proposed to send to President Frank concerning the behavior of the students of the Experimental College and asked of Slichter, "Will you criticize, change, condemn, or approve this statement as a basis for a discussion with President Frank." After three pages of historical background, including the reasons why there had been a conference with Meiklejohn and the house fellows, he continued:

~ From this discussion, several clear-cut points of view developed:

(1) The Experimental College desires as complete segregation and isolation as possible from the rest of the Dormitory students in order to build up a social life about a community of intellectual interests. Mr. Meiklejohn considers anything which accentuates the dissociation of the Experimental group to be advantageous. He even thinks the calling of the men "guinea pigs" has been beneficial in emphasizing this separation and so in developing group consciousness.

(2) Uncouth behavior in the dining rooms is considered evidence of intellectual non-conformity, and so apparently is of no moment, and perhaps by implication a good sign. This would apparently explain why the Experimental Fellows do not attempt to restrain food throwing, excessive noise, etc., in the dining rooms.

(3) The same attitude will explain the disregard of the quiet hour

observance, which has proved disturbing to the non-Experimental houses in Adams and even to Tripp.

It explains the general attitude of disregard of all rules made for the comfort of Dormitory residents in general, and why the Experimental student appears to feel himself above the standards of behavior of the rest of the Dormitory students and unrelated to the organization of the Dormitories. It also explains why the breakage for which damage has been assessed, and which represents only a fraction of the total property damage done, is five times as great in the Experimental houses as in the rest of the houses of the Dormitories.

With these things before us, the Dormitory Committee is keenly conscious of the fact that the original Dormitory project has been very seriously disturbed. As proponents of the original experiment, on which several years of hard work were spent, we are concerned to find that the original experiment has been altered very materially before it has either justified itself or shown itself to be a failure. We are making no criticism of the Experimental College beyond the fact that it has come into the Dormitory set-up as approved by Faculty and Regents, and with views diametrically opposite to that set-up, in several important respects, has jeopardized its success. We no longer have two quadrangles which govern themselves as a unit, whose inmates have common standards of behavior based on mutual consideration and a desire to work out by mutual agreement their social problems; we have instead a family divided against itself, refusing to cooperate, refusing to abide by the majority wish and recognizing no common objectives.

With the entrance of a new class, the Dormitory Association will find itself composed of two equal but entirely different groups; one of which prides itself on its non-conformity and a year's acquisition of habits and precedents which accentuate the division. The other will represent the original conception of the Dormitories and the values they can be made to contribute. This division would not be serious were it not for the fact that the two groups must eat together—they cannot be wholly separated. Furthermore, because the two quadrangles are close together, the quadrangle that recognizes no obligation to respect quiet hours is a decided liability to the men in the adjacent houses of Tripp. Complaints from these residents of Tripp are too numerous to be wholly ignored.

The Committee suggests that the work of the Experimental College might well be carried on in the Dormitories in the spirit in which a transient guest stays in a home—accepting the customs and ideals of that home and fitting in with its general program until such time as he can move into his own domicile. We submit that the major objectives of the Experimental College are not in any way invalidated by a reasonable degree of conformity to the standards of behavior adopted in the Dormitories, and that cooperation in attaining those standards need not interfere with the serious accomplishments of the Experimental College.

We believe that such a reconciliation of objectives should be made at once, and for such time as the Experimental College is to be housed in the present quadrangles. In time it evidently should occupy a quadrangle built specifically for its own needs and with its own dining arrangements, where the desired isolation from the other dormitory students can be secured, and where it can develop its own standards of behavior without interfering with the comfort of others. Until such time, it should so carry on its special problems as to further, rather than to diminish the success of the project which it has joined, and from which for physical reasons it cannot now be separated.

It is our earnest desire therefore that we may meet with you at an early date to suggest such immediate changes as will permit the success of the year's program, and plan for the solution of the problems that confront us for the coming year. We should be glad to have this meeting soon, because the time is near when appointments must be made and policies definitely formulated. We should like also to bespeak enough of your time on this problem so that we can cover it satisfactorily.

Within two days Slichter replied:

~ I have read with interest your letter of March 5. I have gone over it with a great deal of care and I have no suggestions to make, because I think there is no way of approaching this subject except with the utmost frankness and with all the facts before the president. I did not know until a day or two ago of any difficulties of the kind you mention. I have always admired those Englishmen, who in taking up their residence in remote corners of the earth still feel impelled to dress for dinner every evening and to conform to the other conventions that their gentlemen friends at home are adhering to. I quite agree that the success of the Experimental College cannot be bound up in any substantial way with lax manners or a weak recognition of the comfort of others.

This is of particular interest, for Slichter liked to shock people by preposterous statements (often announced in a loud voice) and had the reputation of being an iconoclast. This half-truth could lead to misunderstanding, for fundamentally he was a conservative—even if at times a willful one. The above shows his real attitude when he, himself, was shocked by an outrage against the proprieties.

We have no record as to whether or not Bradley's letter was sent to Frank. There is an oral tradition that at one conference Slichter really exploded at Meiklejohn.

Even as early as 1926 it was proposed to build an automobile road west from the Union along the lake to the dormitories. Bradley's committee, with the willing concurrence of the Regents, was able to defeat this defacement of the campus. Vigilantes have continued to be needed.

The Union. When, in 1927, "The University Committee on the Union" was formed, Slichter was a member and, as in the case of the dormitories, Harold Bradley was chairman. This, however, was not the start of the activity to establish a student union at Wisconsin. The campaign to raise money began in 1919. Sometime later, when the campaign was well along, Slichter gave a short talk on the subject. We do not know the exact date of the occasion, but we have his notes.

~ *The campaign for the Union Building is just 21 years old. It is not surprising therefore that it is now planning to set up in business for itself. The first promoter of this project was President Van Hise. When he was a student here in the 70's, the University community of 350 students was a compact and friendly family with none of the major problems that came later with the enormous growth. President Van Hise's knowledge of the older university of intimate contacts made it natural for him to become the advocate of "a center of community life" as he called the project. Even when I came here in the 80's there were still only about 375 students. Everybody knew almost everybody else. For a number of years I think I knew nearly all of the students in the university. We young tutors were expected to visit our students when they were sick and to bail them out with our own money when they were in trouble. No array of Deans and Deanlets were needed and there seemed to be no need of many rules and regulations. Even our most distant dances and sleigh rides to Sun Prairie and Oregon brought us home at six in the morning, in ample time for the 8 o'clock. We knew each other and we could vouch for each other. Then came the unexpected and enormous growth, and with it came broken contacts and complexity and subdivision. The texture of the university community lost its continuity; it became cellular. A change came over all of us. We found it hard to befriend or to vouch for people we did not know or had never heard of. All of this of course, was in the mind of Mr. Van Hise. He hoped that some of the old friendliness could be restored and that much of the disunion could be done away with through a community* home and commons and the things that would naturally flow there-from. He appealed to the Legislatures for 15 years, but without results. The very growth of the University that made the project a necessity put such heavy demands upon the state purse for operation and maintain-ance, that it set up an obstacle to the solution of the very problem it created. It was a vicious circle. Then in 1919 came forward Mr. Walter Kohler of the Board of Regents. He announced that if we could not get the building from the state we would have to get it from ourselves. With his splendid help the campaign was started. To date 14,000 alumni, students and faculty members have supported the project with contributions. About $1,000,000 in pledges are now in hand. This is the first effort we have ever made to help ourselves. It is a fine expression of our faith that a new era of cohesion and morale is to begin at Wiscon-sin. The project will succeed, first because it is our own job done by our-selves for ourselves; second because it is a memorial to those who went from this university to serve their country, who by the autocracy of war were commanded to leave their fine lives incomplete.

When the Committee on the Union started to function, Slichter wrote, on February 28, 1927, to Bradley:

~ In response to your kind invitation for suggestions concerning the Memorial Union Building and any changes that might be made in its plans at this time, I have the following observations to make:

I think it is obvious that the central feature in making the use of such a building general by the student body is the provision made for taking care of large numbers of students in the commons or restaurant. Stu-dents will probably be able to make more contacts naturally through the dining facilities than in any other way. While this is true, I think a further development of the idea and plan is desirable. The dining to-gether in commons is only one part of the facilities that might be use-fully provided. I suppose in the future there will be many small dining clubs organized in the University; some consisting of faculty members only, some consisting of students only and some more consisting of both faculty and student members. At Cornell University most of the faculty dining clubs now use the Union building and a great many student-faculty clubs have been organized and make use of the facilities given by the Union building for this social purpose. The Union Building plan for Wisconsin, in my opinion, does not recognize this need nor the desirable results that would come from proper facilities. Private din-ing rooms are shown on the plan—the most important of these would be dining rooms that would comfortably take care of from 12 to 20 din-ers. Large private dining rooms that might provide a banquet room for

literary societies or a similar large group do not meet a social need furnished by small dining rooms. Not only that—the small dining rooms
themselves will not take care of the need I have in mind. What is required in addition to small dining rooms are a few small club rooms
furnished with comfortable chairs and a fire-place to which the diners
may resort after the meal for discourse and social intercourse. The small
rooms with dining chairs do not meet this need. It is entirely too un-
comfortable—the big table is in the way and is an obstacle; the soiled
dishes make a mess of the room and there is no spirit of hospitality at
all present. The Madison Club and University Club have made a shocking failure of their attempt to provide for private dining parties. At the
Madison Club after dinner one must resort to the large porch on the
second floor where one is lost and finds difficulty in finding a center for
the party. The University Club has nothing but the basement card
room which would require complete refurnishing to meet these needs.
At the College Club, however, there is a private dining room and the
old library of Colonel Vilas with its comfortable settee and fine chairs
and beautiful fire-place meets this need exactly.

My suggestion is that at least two small rooms which may be designated committee rooms and in which ordinarily committees may meet
be equipped for the purpose I indicate. In order to make them attractive they should be provided with fire-places. They should not be large
rooms but should take care of parties from 12 to 20. If such provision is
made, I believe a number of faculty-student dining clubs or all-student
dining clubs would come into being, and make fine use of these facilities.

I think there is even greater need of consideration of these matters at
Wisconsin than elsewhere. Because of the memorial character of the
building and the necessary formal architecture, the impression given by
the exterior and the interior of the building whatever it may be, will not
be a homey feeling. It will seem, I am afraid, too much like entering
the post-office or the library building. In order to overcome this natural and inevitable handicap special attention and studies should be
given to providing for the real social life of the students in small groups.
The provision for the life of the students in large groups such as dancing parties and so forth is a less crying necessity and of course does not
tend to develop the life of the students in a way the small organizations
may do.

He followed this on March 7, 1927, by a more personal note to
Bradley.

~ Here is a picture of the Committee Rooms in the New Union Building—an expression of distinction and hospitality. I believe the panelling and the fine colonial mantle is the gift of the Union Board. I am told that the furniture and the beautiful rug are the gift of Mr. Bradley, chairman of the Union Committee. The building has no chimney, so the fixtures in the fireplace are of the electric sort—the kind that make a bed of coals you cannot tell from the real sort—like the grates in the lounge of the Hotel York in London, which it took me two weeks to find were not real fire. I think the electrical equipment is the gift of Mr. Slichter. My wife says I am dreaming, but I think not.

The outcome of this is told in the memorandum concerning Slichter written in December 1969, by Porter Butts, who from the opening of the Union until 1968 was its director.

~ . . . His humane view of education, and how it comes about, quickly surfaced and commanded the Committee's attention. At one of the earliest meetings (Committee minutes of March 8, 1927) "a second letter from Dean Slichter was read affirming his belief in the value of small dining rooms with a lounge adjacent (for smoking and leisurely conversation before and after dinner) and submitting a picture and prophecy of how one might look."

The state architect, Arthur Peabody, responded: "The building has so far advanced that it is now largely a matter of taking the rooms as planned and using them to the best advantage."

The Committee thought otherwise. A portion of each of two rooms [now the Beefeaters and the Round Table Rooms] was developed as a lounge area with comfortable upholstered furniture. The floors were changed from terrazzo to wood, carpeted. And a hearth was added, with a gas-fired log—the best that could be done in a building already constructed without chimneys.

Slichter's dining club sought out these particular rooms thereafter; as did most others.

In the same memorandum Butts tells of a bit of advice that may have been more important to the Union than the lounge spaces around hearths with gas logs or electrically simulated fires:

~ My first encounter with Dean Slichter was in 1925, shortly after I graduated and while I was working with John Dollard, secretary of the Memorial Union Building Committee, to develop a new alumni records office for the University and press forward with the campaign to raise funds to build the Union building.

Together we went to Slichter and said: "We thought if we are to be working at the University, and with students and alumni, we ought to keep in touch with what goes on in the classrooms. Is that a good idea? And if so, could you suggest a course? We have no particular career goal in view, at least yet."

"Well, now," he said, leaning far back in his chair, eyes twinkling, "that's all right. Do it. We have a first rate man coming from the University of Goettingen next semester—Oskar Hagen—to develop a department of art history. He'll have a seminar for graduate students. Why don't you think about that?"

So it was that Dollard, a commerce major, and I, an English major—with no exposure to art of any kind, ever, in course or out—found ourselves with one other student in Hagen's first graduate seminar. Slichter had a relaxed attitude about prerequisites and credentials, at least in this instance. Or perhaps he suspected something might come of all this.

As it turned out, something did. With the zeal of a new convert, as I now think back upon it, when I became director-designate of the new Union in 1927 I managed to get a "reception room" (the building was under construction) redesigned to function as an art gallery—the first the University ever had; I undertook to write the first history of art in Wisconsin; we invented the "Wisconsin Salon of Art"; and through an around-the-calendar series of notable exhibitions, lectures, and visitations by famous artists, art became the most conspicuous of the Union's enterprises. I continued to take an art history course or two every term for ten years. I think Dean Slichter looked on approvingly.

Other Committees. Now, in 1971, it is amusing to note that the first university-wide committee of which we have a record of Slichter's membership was a "Committee to Investigate Recent Disorders"—appointed in the fall of 1889. The disorders appear to have been aggravated hazing, involving some shooting but no hitting. Students in court all seemed to testify that it was unethical to "squeal" to the faculty. Some thought that under oath in a court they should "tell all"; others believed that they should not "tell all" except in the case of "grand larceny," but should protect fellow students in the case of "petty larceny."

He was on chore-committees, such as those on timetables, on rooms, on arrangements for examinations, on the university calendar, on reporting absences, and on lectures and convocations. I would put in the same class the committee on honorary degrees, but I guess he really had some interest in this subject.

There were also the chairmanships of the Research and the Graduate Committees when he was Graduate Dean, but these are described in Chapter 6. There were committees that grew out of other committees on which his duties were more central; for instance, because of his position on the Athletic Council and its predecessors, he was on the Advisory Committee of the Northwest League, the Boat Club Committee, the Committee to Consider the Amount of Required Gymnastics, and the Committee on the Professional Course in Physical Education.

The listing of a few other committees will illustrate the range of his services. He was on the Library Hall Committee in 1892 and for some years thereafter—at the time when it was decided to push for a building to be used jointly by the State Historical Society and the University. This committee, as well as a number of others on which Slichter served, was headed by Birge. He served on the "Fish Committee," a curriculum committee of the College of Letters and Science. Another committee, to which he was appointed in 1896, is sufficiently described by its title, "Presidential Regulations on Relations of Student Societies, Fraternities and Young Women of Ladies' Hall [Chadbourne] to the University." In 1897 he served with Frederick Jackson Turner, chairman, on a committee to write a constitution for the intercollegiate debating league. He chaired such diverse committees as that on instruction of artisans in the Army and the one to study the needs for buildings for the College of Letters and Science.

In 1897 Slichter was on a committee concerning cribbing in examinations. It is appropriate here to insert the story of his watching a student for awhile and then announcing in a loud voice, "Bill, if you are looking for your crib it's fallen under the seat."

The Committee on Accrediting of High Schools and Appointments deserves somewhat more attention. (This membership also brought with it being on the Committee on Entrance Requirements.) For many years, ending in 1931, the University was in charge of accrediting the high schools of the State. After the information submitted by a high school was studied, a team was appointed to visit the school and to report on the quality of its work. A faculty committee was in charge of the program, but the visitors were not limited to the membership of this committee. (Professor E. B. Skinner told me how he sometimes served on an inspection team with the ominous names of Slaughter, Skinner, and Coffin.) Slichter's work on this committee culminated in his being chairman in 1908–9. This was one of the committees that he was able to get off by taking a year's leave of absence in 1909–10.

Before he was on the accrediting committee, he had frequently served on inspection teams. He used these visits to make constructive criticisms and to sustain the efforts of the school officers to improve the libraries and the laboratory equipment. He often came to strong conclusions. Occasionally on the basis of these visits he recommended teachers for better jobs. He could also be devastating in his reports. In 1909 he wrote:

~ . . . I am certain that he is the poorest teacher that I have seen in any high school at any time. This I say after visiting hundreds of high schools in the past 30 years.

I should make the points as follows: His manner in the classroom was exceedingly poor. He exasperated and tormented the students and was disrespectful in his criticisms; 2nd, his knowledge of the subject was inadequate. . . . Third: He did not keep track of the work being done by the students on the board, so that he did not know which students were working independently and which were copying from others. Fourth: Innumerable errors were permitted. . . ."

(It was later explained that the trouble had resulted in part from using a combined athletic coach and biology teacher to teach mathematics.) This letter is interesting in several ways. First, it gives evidence of the number of high schools he had visited; but the "past 30 years" takes him back to his own high school days, so perhaps not all of the "hundreds" were inspected by him. Much more important than this illustration of his penchant for sweeping statements is the phrase "disrespectful in his criticisms," for he had an ingrained respect for young people—a respect that was very deep in spite of the fact that on occasion it was successfully camouflaged.

Slichter and Birge were on the committee which, in 1907, strongly defended against the Superintendent of Public Instruction the propriety of a university's inspecting high schools. From the style, I judge the report was written by Birge.

In 1914 there was a notorious survey of the University under the direction of William H. Allen, a professional educational surveyor. This is described in detail in Curti & Carstensen's *History of the University*. Allen was born in 1874; and the history remarks: "For a young man he had traveled far, aided by a capacity for vigorous self-advertisement." The report was considered to be abominable; and the University's reply was formulated by a committee headed by Dean Birge and by Professor George C. Sellery, who told me that he believed his work on this subject did more than anything else to make him Birge's successor as dean when Birge became president. It fell to Slichter's lot to defend the faculty against the charge of laziness. This he did admirably. In his statement he said:

~ *I know from personal knowledge of my colleagues that a university professor ordinarily puts about sixty hours per week upon his university work. Practically all university men must work evenings and if Saturday afternoon is taken off, the loss must be made up by extra hours at other times. This is what is to be expected. University men are men of intellectual ambition and they overwork rather than underwork in their chosen field. The few men who may become drones under the University system do not materially count in the aggregate.*

When this had gone through Sellery's hands the first sentence, as printed, had become: "I believe from personal knowledge of my colleagues that a university professor ordinarily puts from forty to sixty hours per week upon his university work." Sellery was more cautious than Slichter concerning a possible credibility gap.

The Allen report was so unjust and the University's reply so cogent that the State Board of Public Affairs sided with the University and little injury, besides a most unfortunate waste of faculty time, resulted.

The University Club

Over the more than sixty years from its founding to the present, the University Club has had a beneficent influence on the University of Wisconsin. The fact that at this writing it is having financial difficulties, and that the role and nature of the faculty club at the University is in need of clearer definition, does not negate the proposition that without the University Club the University would have been a less useful institution and a less attractive place for faculty members. The Club has furnished a center for discussion, both in the lounge and in the dining room; food of quality, which over the years has varied from mediocre to good (not from poor to excellent, as I started to write); a place in which bachelor faculty members could live (my first year in Wisconsin was spent there); a recreation center—billiards, pool, cards, chess, and table tennis; and an excellent subscription list of magazines and newspapers. If, of itself, it did not break down departmental lines, it gave an opportunity to those who wished to do so.

We have been unable to find the early minutes of the Club, but from the letters written by Slichter we can fairly well conclude what was done. After an earlier abortive effort to form a club of a "mixed" nature (no elucidation of "mixed" is given), a stock company to buy land and erect a building was formed in 1906 or 1907. Subscriptions for stock at $100 a share were solicited by an active group of which Slichter was either chairman or secretary—for additions to a typed list of subscribers were inserted by him. The Club, like most "university" clubs, was open to all persons with a college education, and about half of the money came from nonfaculty members who comprised nearly half the membership. The large majority of the stockholders subscribed for a single share. None subscribed for more than ten, but there were four who subscribed for that number: E. T. Owen, Professor of French, whose home on State Street was on land that ran through to a high wall on Langdon Street—which, after his daughter's marriage to William H. Kiekhofer, was known as "Kiekhofer's wall" and was decorated with murals and slogans by generations of students; John B. Parkinson, for a quarter of a century Vice-President of the University and the person from whom the Club bought the land on which to build, as well as the residence that for some time was part of the structure; Magnus Swenson, who at the time was a Regent from "the State at large" (most regents represented congressional districts); and President Van Hise. Noblesse oblige!

Slichter was particularly active in securing members from outside the faculty, in part because he had a far wider acquaintance in the non-academic world than most of his colleagues. By April 1907, there were subscriptions to nearly $30,000 worth of stock. From the following letter, written June 9, 1908, to Mr. H. L. Doherty of New York, a noted public utilities engineer with Madison connections, we can gather that the enterprise had gone forward rapidly but that more stockholders were needed:

~ *We are just completing at Madison a University Club House which has been erected at the corner of Murray and State Streets. We have put about $52,000 into this enterprise, and expect to open up on the first of February, with a membership of 300. We have raised the money for this Club in stock, about $21,000 among the Faculty of the University and about $21,000 among townspeople of Madison. It occurred to me that you would probably like to become a non-resident member of this Club.*

We have about 22 sleeping rooms in the Club and expect to run a good restaurant, which would enable you to escape Madison hotels when you are in the city. I shall be very glad if you will let us know whether you can help this enterprise by taking some of our stock and becoming a non-resident member. The dues for non-residents are $10 per annum.

Charles Sumner Slichter, about age seventeen

Mary Louise Byrne before her marriage

Slichter as a young faculty member

Dean Slichter

Mrs. Slichter, when her husband was Dean of the Graduate School

The Slichter boys—Louis, Allen, Donald, and Sumner (about 1902)

Dean Kurt Wendt, Donald, Allen, Louis, Sumner, and President
E. B. Fred, Engineering Day, May 3, 1957, when citations were given
to "The Slichter Boys"

The house at 636 North Frances Street, Madison, Wisconsin,
the Slichter home from 1890 until Slichter's death

Sumner Slichter
in the
"nobby and durable
if not fancy"
baby carriage

Two days later he wrote to Mr. O. M. Salisbury of Madison and remarked: "We expect to pay a dividend of 5% after the first year." Sometimes I think Slichter acted on the plausible theory that if you dreamed enough, some dreams would come true. He signed himself: "Chairman, Finance Committee." A year later he was chairman of the House Committee. What other Club offices he held at the time we do not know. Perhaps he was president. He was not secretary, for he addresses Professor M. B. Evans, of the German Department, as such in a letter nominating Chief Justice Winslow to membership in the Club and stating:

~ *Mr. Winslow graduated from the Law School of the University of Wisconsin in '76, and has since received the honorary degree of Doctor of Laws, from the University.*

He is a person of supposed reputable character, and I therefore nominate him for membership in the Club, and trust that his name can be posted on the Club bulletin board at an early date.

(This was not the only membership for which Slichter recommended Winslow, for in 1911 he pushed hard to have Winslow appointed to the United States Supreme Court.)

He was anxious to secure a Chinese steward for the Club and inquired of the Chinese Consul in Chicago about such a possibility. It does not seem to have panned out, but for awhile satisfactory Japanese help was secured.

As the first steward had been "very unsatisfactory," Slichter handed down a decalogue to guide a later one.

~ *1. Inspect each room in the club each day. Note any failure to keep the rooms clean.*

2. Check the sales of cigars each day.

3. Check all laundry and take inventory of linen on 10th of each month.

4. Watch the use of electric light and keep same down to a minimum.

5. Require checks to be signed in advance for all service. This includes meals, billiards, cards, and rooms for transients, as well as dinner parties.

6. In case of trouble with plumbing, telephone
        Anton Metz, plumber
    In case of trouble with steam heater, telephone
        Mueller Company, King St., Madison

In case of trouble with stove or steam table, telephone
    Wolff, Kubly & Hirsig.
7. *Use bills of fare at luncheon and dinner as soon as possible.*
8. *Use napkin rings with names for permanent guests.*
9. *Take inventory of supplies on last day of each month.*
10. *Check all supplies when received and preserve slips for book-keeper. Do not permit merchants to deliver goods without slips. These must bear date of receipt.*

The early success of the Club owed much to Slichter's promotional and managerial work.

As late as December 1931, he still took interest in the Club, protesting the closing of its dining room during vacations and holidays "when those who hold full membership most desire its services."

As far as we know, his membership in the Club was continuous from the time it opened until his death. However, in 1918, when he was helping the war effort in every way he could, he wrote to Walter M. Smith, University Librarian and Treasurer of the University Club, who had reminded him of a somewhat overdue account, a letter which indicated more uncertainty as to what he should do than he usually admitted. He paid the bill, but added: "I hardly know what to do about the club. It does not seem right for me to use funds that way when there are so many pressing needs. Do not continue me on the list after Sept. 30, to which my dues are paid. I will drop out at that time." However, in the Club's files, on a follow-up letter of resignation, is scrawled: "Not accepted." Presumably, for once he was talked out of a decision, but one which he had reached reluctantly.

The Occasional Speaker

It is a great asset to a university to have faculty leaders who can on various occasions be called upon to make a few pithy remarks with the assurance that they will be gracious, witty, and brief. In his day no one in the University surpassed Slichter—although Birge's short talks were excellent and witty, even if less rollicking than Slichter's; and Howard L. Smith was at least as witty, but found it hard not to be sarcastic.

Dinners or other ceremonies to honor Birge occurred frequently: so many decades of service, or becoming president, or retiring from the presidency. Longevity and distinction, especially when combined, must be recognized.

Slichter was toastmaster at the dinner given on March 27, 1915, celebrating Birge's forty years of service to the University. In Chapter

6, where Slichter's administrative colleagues when he became dean are described, there is a quotation from what he said about Birge. We have also manuscripts of some of the other introductions on this occasion—which, of course, is no guarantee that we know exactly what was said. Excerpts from three of these follow:

In introducing ex-President Chamberlin:

~ *It is very appropriate that most of the speakers of this evening bear degrees from this university. They are just some of our own boys. The first of these speakers bears the degree of LLD. Perhaps you may not know what that degree means. You are undoubtedly aware, that, in general, the shorter the course, the higher the degree. LLD. is the degree we give to the short course students in the correspondence school. I well remember when this particular degree was given. We just wrote Professor Chamberlin a letter telling him that if he would come up at Commencement we would give him a degree. We said that there would be no questions asked, no examination. He wired right back that he would come. That is all there is in that kind of a correspondence course. The only difficulty is that the correspondence has to begin at the right end. Otherwise many of us might have acquired the same kind of degree.*

Another speaker was Howard L. Smith, whose speech in the Assembly Chamber, referred to below, was said to have ended: "But do not give up hope, even the Vilas sinner may return." In introducing him Slichter said:

~ *Most cities have had in their history some great calamity from which all their future history is dated. Thus, Chicago dates its events from before and after the great fire. Galveston dates from the great hurricane; Mobile dates from the yellow fever; San Francisco dates from the earthquake; Dayton from the flood. Madison, unlike most cities, dates its happenings from the two great events of momentous importance—a sort of dual standard or historical bimetallism.*

One of these events was the great snow storm which required the dwellers in Madison to tunnel out from their homes and caused Langdon Street and Aristocracy Hill to look like Alpine passes. The other epoch-making event was a famous speech delivered by a young University man in the Assembly Chamber in reply to a speech of Colonel Vilas. This speech set the whole state on edge and set Madison so agog that it has never recovered. When I first came to Madison I found that one's first duty in polite society was to be able to talk about these two great events. One's social position was nothing if he had not acquired facility in passing rapidly from one of these topics to the other or in talking about both calamities at the same time. . . .

Unfortunately, the committee on arrangements cannot reproduce that snow storm for this occasion, but we can and are glad to introduce the famous whirlwind of the Assembly Chamber.

It gives me much pleasure to introduce our own and only, our inimitable and inevitable Howard L. Smith, of the Class of 1881.

The introduction of Dr. Henry B. Faville, of the Class of 1880, included:

~ One need not be surprised at the large number of former students of Dr. Birge that entered the Medical Profession. Anyone who got a whiff of the odors that came forth from the biological laboratories on the upper floor of Science Hall knew the explanation. How any human being could inhale the atmosphere prevailing in those corridors without becoming convinced forthwith that the state of health on the face of the earth was simply rotten is beyond belief. He was simply forced to the conviction that the greatest need in this world was for more doctors of medicine.

To set the tone for the evening he had already declared:

~ The oldest men that I know are about nineteen years of age. At that age a man has usually completely solved all the riddles of existence. You cannot tell him a thing. He has the mysteries of religion and philosophy reduced to a formula,—in fact, the man of nineteen is not only old, but he is fossilized. From the age of twenty on you can usually begin to tell him something. By hard work and constant effort most men are enabled to acquire in forty years a considerable stock of youthfulness.

Slichter could speak with great earnestness, but always accompanied with a bit of fun. What he said concerning "The University of Tomorrow" at the Annual Banquet of the University of Wisconsin Alumni Association, June 10, 1906, follows in full:

~ It is a common saying that all of the great betterments that have come to the human race, all of the great reformations and struggles for liberty have been instigated and carried out by cranks and fanatics. When I think of this great truth, my confidence in the great things that will yet be accomplished by our University Faculty is unbounded. The biggest cranks bring us the best success. It was a very fortunate thing for the State of Wisconsin, when, over twenty-five years ago, the university engaged a man named Henry, a fanatic in agriculture.

It was also a good thing for many of you who are now alumni that our Dean Birge had such crazy ideas about developing something good out of green freshmen. We cannot get too many cranks. I firmly believe that it would be an excellent thing if our Board of Regents, instead of being constituted of sensible business and professional men, could be made up of a lot of cranks. But most fortunate of all crazy ideas is the insane notion of President Van Hise that the University of Wisconsin must become the best university of them all. Some people think that the best way to develop a university is to put on a poultice here and a mustard plaster somewhere else. The faculty have discarded the poultice treatment—they believe nowadays in the open air and sunshine method. We are past the days when we fear too much engineering or too little of the humanities, or underdevelopment in one place and overdevelopment in another. The idea is now manifest that the greatest university of the future must be a university for all the people. While the rudiments of this idea are by no means new, yet it is just now becoming clearly apparent that the University of Wisconsin must exist just as much for the two million people of the state as for the four thousand students on the campus. This idea has so grown within recent years that the one feature which will characterize the present administration is its emphasis of the university as the instrument of the state. The more the President and the Faculty see visions and have dreams of this sort the better for the state and for the University. This development is really nothing but the expression of the true fundamental purpose of education. If I were to characterize in a few words our public educational system, I should say that it is the one human institution that is planned and organized primarily for the needs of the future and not for present needs or present profit. All other human institutions are functions of the present moment. Farms, factories, business corporations, jails and courts of law, are for material and present profit. The public educational system exists primarily for the future needs of the race. We might express this by saying the sensible people of the world are engaged in drawing horizontal lines connecting this thing with that profit or that loss. The fanatics of the University draw vertical lines leading from the things of the present to the more perfect and more spiritual things of the future.

The University has had fifty years in which to make mistakes. I do not know whether or not it has made any mistakes—if it has I am not prepared to admit it. There are two different kinds of mistakes. The mistakes of the University have certainly not been of the irrevocable or nonreversible sort. They are not like the mistake made by the man who was experimenting for the first time with shredded wheat biscuit. This man took the numbered coupons from the package and ate them, served with cream. He carefully laid aside the shredded biscuits and presented them for the set of china given with each package. This is what I mean by a natural, but irrevocable mistake.

I am not going into details about the work of the University for the two million people of the state. Do not think for a moment that the attendance of three or four hundred farmers upon the special two weeks course arranged for them at the University represents anything more than the smallest part of the possibility in this movement. The best example to explain what is in the mind of the President and Faculty is the new movement to develop the scientific courses preparatory to medicine. This, in the plans of the University, means the development of many laboratories, including a laboratory for experimental pathology, and many courses including courses for health officers, etc. The object of all this is not only the training of physicians, but the advancement of science and of the laws of preventive medicine, the dissemination of knowledge concerning infectious diseases and the enforcement of sanitary laws. In the President's biennial report are these words: "At the present time knowledge is available, which, within a decade or at most a score of years, would practically eliminate from the state nearly all infectious diseases. Even tuberculosis could be practically eliminated in one generation." It is a very fortunate thing that the President and Professor Russell and Professor Bardeen have visions, and it is equally fortunate that we have Alumni like Faville, Puls, Ochsner and Dodson to encourage their fanatical plans. If the enthusiasm of these men is not bridled we shall soon be face to face with a situation in which human diseases are given as much attention and study by the state as is at present applied to the diseases of cattle and swine.

One of the greatest fallacies of the day is the criticism by some people of the attention given to research and advanced instruction. These critics would have the future no brighter than the past. It must become the part of good manners not to look lightly upon the visions of our university dreamers. We should no more laugh at our misguided professors than at foolish Benjamin Franklin flying his kites in a thunderstorm while sensible people remained cosily at home. Do not fear that elementary instruction will suffer. The director of our Graduate School has investigated this matter. He finds that those departments doing the most in advanced courses and productive work are the very departments in which elementary instruction is the strongest. He has merely discovered the simple axiom that men that do the most work are the men that find the most things to do.

Only one more point do I desire to make for the university of tomorrow. A few have the idea that a small institution is as good or better for the average freshman as a large university. This is the greatest fallacy of them all. To make the large university the best place for the young student is merely a question of proper organization within the university. The faculty has given hard study to the best management of the elementary courses. The new ideas and new plans that are being constantly perfected are such that no better place for elementary work will anywhere exist than in the University of Wisconsin.

We do not know just when he had the following bit of fun; but it is possible that it was a briefer than requested command performance in compliance with a note from President Adams, dated May 18, 1898, which read: "We must have an eight or ten minutes' speech from you at the Convocation banquet. How would you like: 'Matthew Arnold's Theory of Numbers,' or 'Little Drops of Water versus Little Grains of Sand,' or something else which please name?"

~ It is unfair to ask me to rise before this board of experts. When President Adams first came to Wisconsin from Cornell he soon remarked that Cornell was a singing university but that Wisconsin was a dancing university. We did not contradict President Adams at the time, as we knew that he would soon find out that Wisconsin was only secondarily a dancing university but was primarily a talking university. Now Wisconsin is not my alma mater. I do not inherit from your good mother. I am not one of her children, I am only a servant in the household. It has been my job to wait on you and on the likes of you. I do not inherit the gift of talk. I come not from a singing or a dancing or a talking university, but from a praying university. Your chairman has probably called upon me so that I might pray for you. But that is just what I have been doing for all these long years. I have never ceased, long and loud, to pray for you. I admit that the results sometimes seem very discouraging. I have even prayed for the Regents. I am glad to report that on the whole I can see some results. I am sure I see you slowly edging toward the mourner's bench. I know that I must not believe the impossible. But remember that miracles are only impossible things that actually happen. So I expect to live to hear you all shouting "Halleluia!" and to see you lined up in the mourners' benches, awaiting salvation to descend upon you. When this happens, do not forget—do not fail to remember—that it was I who prayed for you.

He was also asked to write official greetings. The following expresses his ideal, and perhaps that of President Van Hise, of the purpose of the University. (Slichter was the University's delegate to the Drake celebration.)

~ *The President, The Regents, and the Faculty of the University of Wisconsin send greeting and congratulation to the President and the Board of Trustees of Drake University upon the celebration of the twenty-fifth anniversary of the founding of that institution.*

May your University, by its precepts and by its discipline, long continue to serve the cause of learning and of public morals, so that the course of years may record ever increasing service to the people.

Chapter 6 ❧ *The Graduate Deanship*

UNTIL about the time that the United States entered World War I, Slichter, although doing distinguished work in the study of underground waters and having collaborated with T. C. Chamberlin and F. R. Moulton in problems in geophysics and cosmology, played only the role of an intelligent and energetic professor in furthering research. He had in 1905 made a number of suggestions to Dean Comstock as to methods for increasing the enrollment of graduate students. Starting, however, in 1917 and until 1934, when he retired, he was to take a decisive part in developing research at the University of Wisconsin. This phase of his work began some three years before he became Dean of the Graduate School in 1920.

The Research Committee Before 1920

Early in 1917 the National Research Council, which had been founded in 1916 as an offshoot of the National Academy, with the immediate purpose of aiding in the preparation for war but with the longtime objective of increasing the scientific effectiveness of the nation, recommended to a group of leading universities that each create a committee to, as President Van Hise announced to the faculty, ". . . act as a medium of exchange with its sub-committee on Research and Institutions, and to secure additional funds for research." The request was referred to the University Committee which, at the March meeting of the faculty, recommended: "That a committee of five *in addition to the ex officio members* be appointed by the President to act as a University Research Committee and to cooperate with the National Research Council, with power to add to its membership from within or without the faculty; the President and the Dean of the Graduate

139

School to serve as members of this committee *ex officio.*" The faculty approved the recommendation and the President appointed John G. Callan, Mechanical Engineering; J. A. English Eyster, Physiology; Edwin B. Hart, Agricultural Chemistry; Ralph H. Hess, Political Economy; with Slichter as chairman. Hess soon took leave for war work and before fall was replaced by John R. Commons, also of Political Economy. Slichter served as chairman of this committee until he retired in 1934.

It was not long before a project within the purview of the committee came to his attention, a project which had all the elements to engross his interest: military importance, scientific and engineering instrumentation, the leadership of Max Mason, and the participation of his son, Louis.

Late in the spring of 1917, Robert A. Millikan, executive officer of the National Research Council, had called a conference of a group of university physicists to discuss the problem of detection and location of enemy submarines. The problem of determining by sound the direction of a distant ship, even from a stationary base, is difficult. It is far more formidable when the detecting device is on a ship, for, as Slichter wrote Birge, Mason's "task was to put an invention on a destroyer, altho literally vibrating with the noise of steam engines, propellers, blowers, pumps, dynamos, and rushing water, so designed that the device would ignore the noise from all these nearby machines, but at the same time would pick up the faint sounds of a submarine miles away and, in addition thereto, give its exact direction." Mason had ideas as to how this could be done. Loss of time meant both loss of lives and of ships—but there was no quick way to secure funds from federal sources.

The University physicists willingly devoted time to the problem while waiting for some legalized arrangement. Lake Mendota also came free of charge, but the nonsalary costs had to be met. Slichter moved promptly on two fronts. He started to raise money from alumni and, on June 30, 1917, he asked the Regents for help. The following Minutes of the Board describe his appearance before it:

~ *Professor C. S. Slichter appeared before the Board and explained the war research work which is being carried on at the university. He explained that regular donations are coming in for the payment of research assistants, etc., and asked that the Regents appropriate $5000 to aid in the work.*

On motion of President Van Hise, second by Regent Vilas, it was VOTED, *That not to exceed $5000 be set aside from the unassigned* fund of $60,000 in the budget to be devoted to special research relating to the submarine problem, it being understood that Professor Slichter will endeavor to raise a sufficient amount of money for this purpose from private funds, and that this fund may not be drawn upon.

Called vote was taken, all regents present voting and voting "aye."

On motion of Regent Jones, second by Regent Mahoney, it was

VOTED, That the secretary be authorized to act as treasurer of private funds used in the war research fund, and to render such further assistance, if possible, as may be desired by Professor Slichter.

Slichter's campaign to secure private funds from or through alumni, in order to get the program started, was an immediate success. About $7,500 was raised, the largest gift ($5,000) coming from G. M. Dahl and others of the staff of the Chase National Bank. The Gisholt Company of Madison contributed. I have found no complete list of persons solicited, or even of the donors. Among the leaders in raising the fund was William S. Kies, who himself gave $300 and later helped organize the Wisconsin Alumni Research Foundation.

Slichter also interceded through the National Research Council to keep the draft from interfering with this work.

In about a month the program was moved to New London, Connecticut, and later came under the Naval Experiment Station organized there in 1917. With the support of Commander Clyde S. McDowell, who courageously cut red tape, the work progressed successfully. In 1921, doubtless upon the initiation of Dean Slichter, the University conferred the honorary degree of Doctor of Science upon the Commander. Slichter had suggested that both Commander McDowell and Professor Mason receive honorary degrees at the same ceremony, but Mason had to wait for his until 1926, when he was President of the University of Chicago.

Louis Slichter was Mason's right-hand man in this work, and before the end of the war both he and Mason were abroad, supervising and participating in the installation of instruments on ships and the testing of them in the danger zone. Louis's father bent backwards, informing him that he was receiving less help from the special fund than others did, but he was also willing to personally make up the deficiency. When Louis was elected to the National Academy in 1944, Mason congratulated the Dean and took occasion to speak of the great significance of Louis's antisubmarine work in the second World War.

Slichter's instinctive recognition of youthful ability was illustrated when he was struck by the competence of a young naval officer he met on a train. Inquiry turned up the name—Chester Nimitz.

"*Number 9.*"   Another prelude to Slichter's deanship was the creation of a special research fund. In the years immediately preceding the
appointment of the Research Committee there had been considerable
progress in securing fluid funds for research. In May 1915, Van Hise
appointed a special committee on the improvement of graduate work
and research. On February 21, 1916, Professor L. R. Jones, Chairman
of this committee, reported to the faculty recommending among other
things that: "The faculty respectfully recommends that the Regents
make additional provision in the budget of the University for the
encouragement and support of research of larger or more specific
character and for the publication of the results of research in general."
The Regents at once implemented this recommendation to a modest
extent, but it was not until May 5, 1919, that President Birge was able
to announce that the appropriation bill "contains for the first time a
definite appropriation for research amounting to $23,000 annually."
He suggested that the initial use of the fund be for projects which
could yield quick results and hence help insure continuation of such an
appropriation. The number of this item in the State Budget was 9 and
for years the fund was fondly known as "Number 9."

On May 21, 1919, Slichter as chairman of the Research Committee
sent the first notice to the faculty that requests for research funds were
in order. It marked an event and described a procedure of great import
to the University. The text follows in full.

~ *To Members of the Faculty:*

*By appropriation of the State Legislature there has become avail-
able an annual sum of $23,000 for the promotion of research in the
University. A part of the fund has been allotted for certain major
projects, but a considerable portion remains available. The President of
the University has requested the Research Committee to canvass the
faculty for information concerning projects of research for which
special aid is needed and for which funds, or adequate funds, are not
available from any other source. Applications will be received and considered by the committee until June 10, 1919. It is believed that allotments can be authorized by the President of the University before the
beginning of the summer vacation.*

*Enclosed herewith is a suggested outline of points which it is
desirable to cover with some degree of fulness in making application
for grants from the present fund.*

UNIVERSITY RESEARCH COMMITTEE
Charles S. Slichter, Chairman
253 University Hall

The outline which followed set a pattern that has become a stereotype, including subject and purpose of the investigation, its present status, the personnel involved, the probable cost, the estimated length of time required, and possible future developments.

The first allotments consisted of three major items—two professorships and the support of Professor Guyer's research, and, in addition, an allocation of $4,870 for eleven grants for ten departments. Among these grants was one which initiated the position of glassblower in the University.

The above indicates the usefulness of this fund, but it had an additional value not discernable from the actual expenditures, since a fund controlled by a small committee, especially a committee with an able chairman who enjoyed its confidence, can quickly be used to underwrite meritorious projects that preferably should be supported from other sources—sources that could only be tapped by more glacial processes. Thus at the start, it was announced that because of this fund, two new professorships, essentially for research, one in physics and one in plant science, could be created. Later, after considerable negotiations and a small expenditure from the research fund, these professorships were supported from other portions of the budget. It was possible to open on the single knock of opportunity, and it was important to the University to bring back Max Mason and to secure Ezra Kraus. Moreover, the approval, through a grant, of one's colleagues who served on the committee meant much to faculty members.

Some chairmen thought of their departments in the European tradition, as satellites to themselves, and chiefly reported their own needs. This attitude diminished with the growth of the University and under the influence of the Research Committee headed by Slichter.

Appointed Dean

In the spring of 1920, Dean Comstock, age sixty-five, requested that he be relieved of his duties as Dean of the Graduate School.

With Slichter's long record of stimulating able students and representing the University with both verve and distinction, and with the proof just before him that the Chairman of the Research Committee could raise money imaginatively and distribute it wisely, Birge could have had few qualms when on June 14, 1920, he wrote the Regents:

~ *To Each Regent:*

I send this note to tell you that I am proposing to nominate Professor C. S. Slichter to succeed Professor Comstock as dean of the graduate school.

For some time I have been consulting with the graduate committee and with other members of the faculty regarding the filling of this position. Professor Slichter goes in with the full concurrence of all of the men whom I have consulted. He is already chairman of the research committee, and in that position he has had much knowledge of the graduate work. He will make it one of his first functions as dean to get the University into as close association as possible with the National Research Council and with other agencies for research throughout the country. This he will do in addition to the standard work of the dean of the graduate school.

No man in the faculty is more especially or more actively interested in the promotion and development of higher study and research than is Professor Slichter, and I feel that the University is fortunate in finding him willing to take this important position.

Birge announced the appointment at the faculty meeting on June 21, and the Regents approved it by approving the budget which included Slichter's name as Dean of the Graduate School.

It seems quite possible that Van Hise in consultation with Birge had thought of Slichter as a possible successor to Comstock when he appointed him Chairman of the Research Committee. As early as 1904 he had received from Slichter a letter which included:

~ *The first essential in the bringing about of the elevation of university ideals along this line is the selection and promotion of the instructional force upon a basis of research ability. It goes absolutely without saying that ability to teach should be required of all. Such ability is very common, but only a small number of persons have the enthusiasm, energy, and capacity to conduct profitable lines of research. For this reason I suggest that no promotion of any kind should ordinarily be made except upon the basis of published writings.*

In spite of this dictum Slichter could easily find that conditions were not "ordinary" when it came to promoting a man with ideas but a meagre publication record.

Certainly placing "Number 9" under a committee of which Slichter was chairman rather than in the various colleges presaged his appointment as dean. It also set a precedent which was to be of great importance when the Wisconsin Alumni Research Foundation was developed.

Slichter inherited the duties of Dean Comstock in regard to students and curriculum but brought with him the function of stimulating and financing research.

Administrative Colleagues

When Slichter became dean, his fellow deans and the president were all colleagues of many years. With most of them he had been closely associated. In spite of his thirty-five years on the campus, two of them had arrived before he did. Birge had joined the faculty in 1875; and Russell had come as a student in 1884, had been in Birge's classes, and become a member of the faculty in 1893.

President Birge, who has already entered into this story at several points, was born in Troy, New York, in 1851, studied at Harvard from 1873 to 1875, and received his Ph.D. from that university in 1878, after he had joined the faculty of the University of Wisconsin. He had been dean of the College of Letters and Science from 1891 until 1918, when he succeeded Van Hise as President.

For many years Birge and Slichter had been fellow members of Town and Gown and of the Madison Literary Club, organizations described in Chapter 8. Slichter had spoken at several of the numerous celebrations of Birge's long service. On March 27, 1915, at a dinner in honor of Birge's forty years of service to the University, an occasion already described, he had said:

~ *One of the most vivid impressions of my life is the first faculty meeting that I attended, not so many years ago, in the old North Chapel in University Hall. John Bascom, of blessed memory, presided. I well remember when he turned his eyes upon the little row of tutors on the back benches, that I never came so near wanting to behave myself in my life. It was a real faculty of real men. Besides seven or eight who are here tonight, there were Professors Allen, Freeman, Frankenburger, Heritage, Davies, Daniells, Bull, Rosenstengel, Stearns, Irving. All of these, in their lifetime, were friends of Dean Birge. Tonight, let it be understood, we speak their friendship as well as our own.*

One black-eyed fellow at that faculty meeting especially attracted my attention. I learned in whispers that he was the Professor of Zoology. Later I found that the department of Zoology was a mere side issue with him,—that his real jobs were very different. One of his real jobs, for example, was to teach the Bible Class at the Congregational Church. Another was to lead a Shakespeare Club that met about town. Another real job was to read every week a wheel-barrow of the latest novels that an Irishman regularly wheeled to his house. Judge Bunn used to insist that Dr. Birge's real job was his attempt to become the best whist player in Madison. But, in addition to these and many other real jobs, the major job of all was that of nurse-maid—a job that he has always held—I mean nurse-maid to that squalling infant, the University. Let me read from a letter from our former colleague, Professor Charles H. Haskins, of Harvard.

"In my early days at the University of Wisconsin I remember Dean Birge's telling me in a conversation that I had not sat up nights with the University as he had. Dean Birge did indeed sit up nights with the University in the days of its youth, and his devotion deserves fullest recognition in the days of the University's prosperity."

Think of that nurse-maid, rocking the cradle while others slept, administering paregoric and peppermint while others were on vacation, untiring, uncomplaining, and everlastingly on the job, and you have a fair picture of Dean Birge. The University grows big and bouncing, but it will take generations to become mature; it is still but an infant. It gets the colic just as hard and just as often as it ever did. I do not suppose that colic ever killed a baby, but when this big infant gets it now, its howl seems to waken the whole neighborhood. I only wish that all of the baby's well meaning aunts and grandmothers—such as the state authorities and state legislatures—would not attempt of themselves to readjust the safety-pins on its inner garments without the professional advice from our faithful nurse-maid.

On February 2, 1920, Slichter wrote to Birge, congratulating him on being able to increase faculty salaries. The letter merits quoting in full:

~ My Dear Birge:

Your announcement of increased standard of pay for University Faculty is a great triumph for your administration and will greatly strengthen the University in its ability to get and to hold strong men. However, those of us who have been many years at the University are not entitled to receive one cent more salary unless the University compensates properly the new President of the University. The Regents, I know, must realize, in a way, your services to the University. It would not be surprising, however, if they were not conscious of the great service you bring to the institution. You contribute not only your own ample good sense and good judgement, but in a most fortunate way are able to bring down to us the best that is now applicable from the administrations of Bascom, Chamberlin, Adams and Van Hise. It is a most remarkable fact that the University has in control the services of such a president and we older men of the faculty should quietly stand back until you are first amply rewarded financially.

I do not think that there are many people who actually know how

*much your good sense and good judgement and untiring cooperation
contributed to the success of the administrations of Bascom, Chamberlin, Adams, and Van Hise. I do know, however, and I know that with
the best intentions of the Regents, you are hardly likely to get all of the
back pay actually due you.*

Sincerely,
Slichter

It is interesting to note that Birge's salary as president was $10,000
per annum. This, however, was continued as a pension until 1932,
when it was reduced to $6,000 and became subject to the same modi-
fications ("waivers") as other university salaries during the depression.
He also received a "Carnegie" pension of $4,000 per year. He made
the most of these by living to be nearly ninety-nine.

Birge and Slichter had much in common. In addition to being able
scientists, both were humanists with a deep love of literature. Both
were brilliant essayists and enjoyed discussing the papers presented to
"Mad Lit," and both were wits of the first order. Each had a deep
respect for the other.

The two senior deans were Harry S. Richards of the Law School and
Frederick E. Turneaure of Engineering.

Richards who was born in Iowa in 1868 had graduated from the
Harvard Law School in 1895, after having received his undergraduate
degree from the University of Iowa. He had been a member of the
faculty at Iowa from 1898 to 1903, when he came to the University of
Wisconsin as Dean of the Law School. He was able, quiet, at times
sardonic, with a penchant for puns. He died in 1929. He was a respected
and well-liked member of the faculty. As far as the records show, his
official contacts with Slichter were not great, but they were near and
congenial neighbors in their places on the north shore of Lake Mendota, and Richards's widow and sister-in-law remained life-long friends
of the Slichter family.

Turneaure was born in Freeport, Illinois, on July 30, 1866. He received his degree in civil engineering from Cornell in 1889 and came
to Wisconsin as professor in 1892, was Dean of the College of Engineering from 1903 to 1937, and died in 1951. Thus he was in office during
the whole period of Slichter's deanship. Turneaure was an able scholar
noted for his pioneering work in reinforced concrete, a trusted and
skilled administrator, but not an aggressive leader. He and Slichter
worked on committees together both before and after Slichter became
dean, and their professional association was close since Slichter's scientific works had a strong engineering slant.

As in the case of Richards and Turneaure in 1903, Harry L. Russell and Charles R. Bardeen had become deans in the same year, 1907.

Russell was born in Poynette, Wisconsin, on March 12, 1866. He graduated from the University of Wisconsin in 1888 and received his Ph.D. from Johns Hopkins in 1892. After a year as fellow at the University of Chicago, he returned to Wisconsin as an Assistant Professor of Bacteriology in 1893, and became Dean of the College of Agriculture in 1907. He and Slichter had known each other well before Slichter became dean, but their most intimate administrative relationship was to occur in connection with the development of the Wisconsin Alumni Research Foundation. Russell was a man of great stature, both physically and intellectually. He was a powerful ally and a fearsome foe. People said that he believed any good proposal would be improved by one administrative "No"—an unfair accusation but typical of his reputation. Halsten J. Thorkelson, a member of the engineering faculty, who acted as the business officer of the University from 1914 to 1921, told me the following anecdote. He, Russell, and another faculty member went camping in northern Wisconsin. A local minister visited campers and pestered them concerning their souls. Soon after they arrived at their camp, Thorkelson started to town for supplies and met the evangel, who asked whether anyone was in camp. Thorkelson replied, tapping his head, "Yes, but he is a bit dangerous." As he returned to camp Thorkelson again met the minister. He had glimpsed Russell —all six-foot-two and 240 pounds of him—with a red bandana tied around his massive head and an axe in his hand. One sight was enough. The saving of souls had been postponed.

Russell, outside his office, was an active member of the Madison Literary Club, an ardent traveler with an ever-curious mind, and an interesting companion. He and Slichter could enjoy each other's company, respect each other's firmness, and cooperate to the benefit of the University. An admirable life of Russell by Edward H. Beardsley has recently been published by the University Press under the title, *Harry L. Russell and Agricultural Science in Wisconsin.*

Bardeen was also a "character." He was born in Kalamazoo, Michigan, in 1871, did his undergraduate work at Harvard and his medical work at Johns Hopkins, where he received the M.D. degree in 1897. He came to Wisconsin as Professor of Anatomy in 1904, and became the first dean of the Medical School in 1907. His rather slow and grumpy manner belied both his mind and his nature. One rarely saw him without a pipe in his mouth, from around which came sounds resembling speech. There was humor in his makeup, as when he introduced Dr. Waters, the anesthetist, to the faculty as the only one of its members explicitly hired to put people to sleep. Moreover, he was capable of swift action. Warren Mead of the Department of Geology told of the occasion when another department—I believe physics—moved out of Science Hall. The geologists, believing that possession's nine points were of value, decided that early the next morning they would occupy some coveted space. Upon arrival they were astonished to find Anatomy already ensconced. The best they could do was to compose verses, each of which ended with, "You're a better man than I am, Dean Bardeen." Bardeen remained dean until his death in 1935, one year after Slichter had left the deanship. We have few references of any contact between the two men, but they must have collaborated extensively, for Slichter was influencial in getting support for various projects in medical research.

The oldest of the deans in age, but not in service, was Louis E. Reber of the Extension Division. He was born in Nittany, Pennsylvania, in 1858. He graduated from Pennsylvania State College in 1880 and later studied at Massachusetts Institute of Technology. He came to Wisconsin as Director of Extension in 1907 and became dean in 1910, resigning in 1926. He died in 1929. He did much to put extension work on a sound basis. Slichter's urging that the Extension budget be cut in greater proportion than other budgets to meet the decrease in funds during the depression came after Chester Snell had become dean of Extension.

The junior dean, both in age and in length of service as dean, was George C. Sellery, born in Kincardine, Ontario, in 1872. He received his B.A. from Toronto in 1897 and his Ph.D. in history from the University of Chicago in 1901, in which year he came to Wisconsin as an instructor. When Birge became president, he chose Sellery to be his successor as dean of the College of Letters and Science. Sellery was distinguished both for his scholarship and the excellence of his teaching. Although frequently disagreeing and sometimes annoying one another, Slichter and Sellery were each the kind that enjoyed the quality of mind and the robust humor of the other. They cooperated in bringing strong men to the faculty, but their arguments sometimes required presidential mediation.

The following certificate bears witness to the fun they could have with each other.

~ WHEREAS *Professor Charles Sumner Slichter, Esq., M.S., Sigma Chi, Phi Beta Kappa, Sigma Xi, etc., affirms: first, that he saw a robin on the afternoon of Wednesday, February 7, 1912, in the vicinity of his shack; and, secondly, that no one would believe him if he said so.*

Now THEREFORE, we, the undersigned, do swear, jointly and severally, with "solemnity and moderation," and with full knowledge of the responsibility we are assuming, that we were with the said Professor Charles Sumner Slichter, etc., on the afternoon in question. And we do further swear that we were with him, at his invitation, to celebrate "election day," and that at about two o'clock in the afternoon of the said day, that is, about one hour before lunch, we saw him see the said robin.

IN TESTIMONY WHEREOF, we have hereunto affixed our hands, in the presence of Mr. William Dixon Hiestand, Registrar of the University, Secretary of the Faculty, and a Notary Public for Wisconsin whose lease has not expired, this fourteenth day of February, in the year of Our Lord one thousand nine hundred and twelve, and of the independence of America the one hundred and thirty-sixth.

> Dana Carleton Munro
> G. C. Sellery

Congenial disputants, they still differed fundamentally. When I succeeded Sellery, he remarked to me, "I believe I've done more harm to the University by the 'Yeses' I have said than by the 'Noes.'" Slichter would not have worried about the harm he had done, but would have been elated by the results of many of his enthusiasms. Farragut at Mobile Bay might well have been one of his heroes.

These were his administrative companions. He liked them and they liked him. He also enjoyed crossing words with them, for they were worthy of his metal.

The remainder of this chapter describes, often by illustration, his work as dean. A charge against it, of lack of unity, could be sustained; but lack of unity is a characteristic of the tasks that a dean performs.

Relations with Students

Few equally successful administrators have been as unmethodical or as unfettered by precedent as Slichter. This does not mean that he cared only for policy and would not be bothered by details. Rather, he involved himself in the problems of many students, but some could not discern from his actions what policies guided him. I am not sure he could. Many were amazed at the correctness of his decisions which often seemed almost to be based on ESP. His own explanations at times were irrelevant.

Anecdotes are in order.

A student from the University of Michigan, whose name I do not know, was appointed an assistant at the University of Wisconsin in a strong department, whose name I will not divulge. When he came to register in the Graduate School, Slichter refused to admit him; when the student asked why, the response was, "You got a C." The student— "Can't you get a C and be admitted?" Reply, "Not from Michigan." The department was outraged and urged—in these days we might say demanded—that he be admitted. Slichter refused and the student left the campus. A little later the departmental chairman told me, almost with awe, that the student had been convicted of a major crime and was serving a sentence of several years in another state. Was Slichter protecting a secret, or did he just size up the young man as one whom he did not want around? Most of us guessed the latter.

A visiting teacher in the Mathematics Department who wanted to take some graduate work tried to persuade Slichter to make an exception to some regulation or other, and he abruptly refused permission. As she left the room, her Irish temper rose; she turned around, walked back, and said, "All I want to say is that you can't make me cry as you do the young things." The next time they met was at a dinner of the Mathematics Department, and he asked for the privilege of escorting her—a wise choice on his part, for both Theo Donnelly's wit and the stories she told were fascinating. I hope she told him the story of the freckle-faced little boy in the football suit to whom she gave a lift, and who declared that "Holy Rosy's going to knock the stuffin' out of Immaculate Conception."

It was often stated that some days were better than others for making a request from the Dean. There was no doubt that his mood varied and that he exploited rather than hid these variations. I question, however, whether many decisions were altered by these moods.

He hated sham or educational trivia and he liked the picturesque in speech. He sought quality and would smooth the road of those with real ideas. Thus the student from another university, who presented "whittling" (presumably the transcript used another title), was refused credit unless he had the concomitant course in "long distance and accurate spitting"; but a brilliant young mathematical physicist, who wanted to get on with his work rather than see to the typing and editing of his thesis prior to getting his degree, was told to tie up his notes with string and give them to Slichter in lieu of the manuscript.

Slichter cared a great deal for justice. At times he could explode, but he also knew when to handle a case without fanfare. Once I was asked by him to read a Ph.D. thesis and report personally to him about it. The candidate, upon advice of his major professor, had window-dressed the dissertation with irrelevant and awkwardly used statistical formulae. I reported that the thesis was below par but not hopeless, and that the student, though not brilliant, was not completely dumb and was far less at fault than the professor. I advised the Dean to grant the degree as recommended by the reading committee. This was done; but I have often wondered whether Slichter had anything to do with the fact that the professor soon left for "a better position." For a man who shouted from the housetops Slichter could be very secretive.

Even if exceptions in favor of ability or to satisfy the demands of fairness were made, there was a constant push to raise standards. The following is a letter to members of the graduate faculty.

~ If the standard for the doctorate can be materially raised, the general situation would be greatly improved. The members of the Graduate Committee agree that about twenty of the degrees granted last year at Wisconsin (out of a total of 139) were granted to persons of distinctly low ability. The early elimination of these candidates would constitute a real service to academic scholarship. In some departments of the University the Master's examination is used as a preliminary examination, and incompetent students are eliminated at that time. In any case, it seems important to give the preliminary or qualifying examination at an earlier date than is now held in many departments.

The University of Wisconsin has peculiar responsibilities in holding to high standards for the doctorate, as we grant many more doctor's degrees than any other state University. The surest way to eliminate the unfit is through an early qualifying examination. Some members of the staff seem to have the impression that the preliminary examination is an attainment examination planned to test the knowledge of the student in special fields. This is not intended to be the case. The preliminary examination is intended to test the mental qualities of the student, the passing of which should admit the student to candidacy for the doctorate. It can well be given at the end of the first year of residence or early in the second year.

Of necessity, Slichter delegated much to his secretaries and indeed allowed some of them to become adept at forging his signature, which boosted their egos. On one occasion a student mentioned planning a certain course of action just to have the Dean say that it was impossible. "But you have given me permission," and the student showed the written permit with Slichter's presumptive signature. As was frequently the case, it was impossible not to overhear what was being said, and there was much perturbation in the outer office; but the Dean merely remarked, "I must have been crazy that day." Many have testified that he could "raise hell" in the office, but he would not do it when a staff member used the authority given to her. If a mistake were made, it became his mistake, even the trite phrase he himself would not have used; however, there was a real ruckus when the desk he retained as a consultant after retiring, was cleared and moved. He must also have delegated proofreading, and we are not sure who allowed to pass his request for the Regents to accept a grant "for persecution of projects."

Slichter was worried about "all work and no play" on the part of graduate students. At the first reception and dinner for graduate students at the Union, he was asked to say a few words of welcome. He proclaimed, "The trouble with you graduate students is you work too much. Now that isn't going to make you a human being, the kind we want you to be. You ought to have some fun. So come over to this building right along, and play a little. You'll be a better graduate student."

The same principle carried over to attractive young secretaries, one of whom, Rosetta Powers, many years later (in the letter already quoted in Chapter 3) reported, "On a beautiful spring day we had been working on a report for the regents or something which irritated him no end —about 2:30 P.M. he slammed the books shut and said to me, 'To hell with it—You go out and gambol on the green. You don't belong cooped up in here.' "

Perhaps this should be contrasted with his declaration that the accrediting associations wasted a lot of time, for he could examine an institution in the evening—if the lights were on in the libraries and laboratories at midnight it was all right; if the buildings were all dark, the institution was no good.

He helped protect the time of graduate students in other ways. The following is self-explanatory:

July 19, 1933

~ Governor A. G. Schmedeman
State Capitol
Madison, Wisconsin

My dear Governor:
    I desire to call your attention to a bill, No. 825A, now in your hands for consideration, which provides that every able-bodied male student in the University shall, during his first and second year of attendance, receive instruction and discipline in military tactics.
    I desire at this time to call your attention to the fact that this bill, if approved and made a law of the state, would result in killing the graduate and professional schools at the University. I do not suppose it was the intention of the framers of this law to bring about such a catastrophe, but certainly Law, Medical, and Graduate students could

*not properly perform their work under such a statute, and I am quite
sure there is nowhere in the United States a regulation of this kind now
in force.*

Very respectfully yours,

The catastrophe was averted.

Financial Support of Graduate Students. Slichter was concerned
about the financial problems of graduate students. Starting as chairman
of the Research Committee before he was made dean, and continuing
throughout his deanship, he campaigned for more scholarships, fellowships, and research assistantships. With the growth of funds from the
Wisconsin Alumni Research Foundation, he was eminently successful.
When in the thirties the depression made it difficult for those receiving graduate degrees to find jobs, he played a leading part, as described
later, in the move to give special research appointments to some
twenty-three new doctors of philosophy for the year 1932–33.

When he became Dean, there was a fifty dollar deposit required of
those receiving the Ph.D., to be forfeited if a thesis was not published
within a year. He carried on a vendetta against the strict interpretations
of the business office, especially when the thesis had been accepted for
publication but had not actually been published within the year. During the depression he succeeded in getting the University to accept the
personal notes of students for the fifty dollars, rather than requiring
cash. At the time not even Slichter saw how ridiculous this fee was, and
how beneficent it would be to leave many theses unpublished. Wisconsin was relatively enlightened, for those were the days when Colum-
bia University would not grant the degree until the thesis was published. At present there is a fee of just over $20.00 which pays for
microfilm copies of theses for deposit in the library and for the printing of abstracts of 600 words or less. This, for those whose theses are
not quickly published, is one of the rare decreases of the cost connected
with education.

Slichter spent much effort to secure financial support for graduate
students and to make sure that those the University employed were
not exploited.

During his deanship the enrollment in the Graduate School steadily
climbed until 1931–32, when the effect of the depression became apparent and there was a sharp decrease in the number of students, both
graduate and undergraduate. The proportion of graduate students who
received some form of financial support from the University is shown
in the following table which may contain minor errors (in particular, a

few research assistants may have been classified as teaching assistants when they were working on the so-called "organized research").

	1920–21	1931–32	1933–34
Number of graduate students	570	1384	1034
Percentage of fellows	9	9	9
Percentage of scholars	5	3	3
Percentage of assistants (chiefly teaching)	37	30	30
Percentage of research assistants		10	10
Percentage on University appointments	51	52	52

The teaching assistants were not, and are not, budgeted or controlled through the Graduate School. The source of funds for the other three categories was chiefly in the Graduate School, and the proportion of students in these groups rose from 15% to 21% during the period from 1920 to 1931, and was maintained at that level during the depression. Also, some graduate students must have been employed by the University as student help on an hourly basis, without title. It is true that, in 1920–21, a considerable number of the 318 instructors were graduate students, while a smaller proportion probably was in the later years.

Thus the big advance was in the building up of a program of research assistants working on faculty sponsored projects which often provided material for theses. More than any other person, Slichter had developed the pattern for this program. The chief source of funds for this purpose, at least during the later years of his deanship, was the Wisconsin Alumni Research Foundation. The start had been made from grants from "Fund No. 9."

The cost of living, unbelievable as it may seem to the present generation, decreased by some 25% from 1920 to 1934. However, it had just reached a peak at the start of this period, while salaries and stipends had not kept pace. Catching up was needed, and the stipend for fellowships was raised from a standard of about $500 for the academic year in 1920–21 to $600 by the end of Slichter's deanship.

Slichter was active in seeing that half-time assistants were not given such heavy teaching loads that they could not do justice to their graduate work. Although teaching assistants were budgeted in the various colleges, it was to Slichter that the Biology Division turned in the winter of 1920–21 to champion this cause.

Slichter rightly believed that it was a good thing to have a group of first-rate graduate students from outside the United States and was active in arranging to help them financially. There were already a few funds in operation, such as the Commonwealth Fund, which helped foreign students in many institutions; but special funds explicitly for students at the University of Wisconsin were lacking, and the ethnic pride of citizens descended from the various national groups was appealed to, to furnish them. Moreover, he strove to keep the out-of-state tuition from rising substantially above that of competing universities.

In asking support for Swiss students, Slichter not only appealed to this national pride but trotted out his own Swiss ancestry. In 1927 he wrote to a banker of Swiss descent:

~ A matter of very great interest to Swiss-American citizens of Wisconsin came to my office early this year from the President of the Swiss Universities. It seems that Professor Habicht, a prominent Swiss professor of Harvard University, visited Switzerland a year ago and it was arranged with the presidents of the Swiss Universities to establish ten scholarships for American students at the Swiss universities and for ten scholarships for Swiss students to study at American universities. Professor Habicht wrote me nearly a year ago and, as both my father and mother were Swiss, I took a special interest in this project, and have been working to have one of these Swiss Scholars appointed to Wisconsin and one Wisconsin student to be received at a Swiss University. This has now been made possible and a very fine young man is now here from Switzerland. I have talked this matter over several times with Tom Hefty and was glad to find that he agreed with me that we should not pass up this arrangement.

I must now provide for the support of this scholarship. We know of you at the University as a very prominent and loyal Swiss-American and I write this letter to interest you in this project and to ask you to support the scholarship at Wisconsin for at least three years. It will cost $1000 each year, and in exchange for this the Swiss Universities will support free at a Swiss University a student from Wisconsin. It is difficult to give a good account of the matter in a letter, so Mr. Hefty and myself would like to drive down to Monroe at your convenience and talk the matter over fully with you if you will kindly set a date.

As you undoubtedly know, there are no finer universities in Europe than the Swiss universities and there are no finer people over there than the Swiss people. I would want to carry out our half of the program even though my family pride did not give me a special right to be interested in the matter.

As far as we can find out, this was not his most successful venture in fund raising. He could not have fared worse if his ancestors had been Scots.

It was in connection with fellowships for foreign students that he wrote President Frank:

~ Although we have been able to secure so many foreign fellowships for our graduate students and have been so successful in securing more than our share of these fellowships supported by outside sources, I shall feel rather cheap if we do not contribute a little bit from our own funds. I am glad to stand up to the free lunch counter and take my share, but my conscience troubles me when I dump all the olives in my pocket.

In the first decade of Slichter's deanship, the number of foreign students in the Graduate School increased by over 50%, while the total of foreign students outside the Graduate School fell appreciably. The number of fellows and scholars from outside the United States nearly tripled. The depression, of course, decreased the enrollment of foreign as well as of American students.

When Slichter became Dean of the Graduate School, the University had granted 1,853 master's degrees and 442 degrees of Doctor of Philosophy (the first being to Charles R. Van Hise). During his deanship, 4,481 persons received the master's degree and 1,350 the degree of Doctor of Philosophy. In the first year of his deanship 191 persons received the master's degree and 39 the Ph.D., while the corresponding figures for 1933–34, his last year as dean, were 266 and 126. Then, as now, Wisconsin was among the first four or five institutions in the number of Ph.D.'s granted, and in certain fields played a dominant role.

Younger Faculty Members

The generation gap was almost unknown to Slichter, and when there was one it was between him and others of an age similar to his, but whose outlook was less youthful.

His ability both to share and to elicit the enthusiasm of youth was shown long before his deanship, in his relationship with his sons as well as with other young relatives, and with his students. As dean, the same genius came into play in dealing with younger faculty members. This cannot be described in general terms for it was an individual matter in each case. I shall give a number of illustrations.

The first four—Warren Mead, Warren Weaver, Karl Paul Link, and Perry Wilson—all later became members of the National Academy, a fact which not only is to some degree a tribute to Slichter's influence, but also shows that he enjoyed the company of able minds.

Warren J. Mead, a geologist who graduated from the University of

Wisconsin in 1906, was for many years in frequent contact with Slichter, receiving from him stimulation and understanding (not only of his science, but also of his humor) and giving to him unbounded admiration. It is reported that in connection with the requirements for the doctor's degree, Slichter accepted the word of geologists, rather than linguists, in regard to Mead's proficiency in a foreign language.

Mead was a gifted photographer and took a number of pictures of Slichter as dean, one of which appears among the illustrations.

When, about 1933, Mead was considering going to the Massachusetts Institute of Technology, many were concerned, and some of us went to Slichter to try to get him to do all in his power to hold Mead at Wisconsin. Slichter was already trying, and deeply wished to keep Mead, but he realized better than we did that the University would not fall apart if Mead left, and finally exclaimed, "Don't worry so much; if we can have a man for the twenty years he is making a reputation, let someone else pay for it." Slichter was willing to gamble on youth even in comparison with an admired and admiring friend.

Warren Weaver's relation with Slichter was extremely close. He was in Slichter's class in freshman engineering mathematics in 1912, took his Ph.D. under Max Mason, another student of Slichter's whose relationship with Slichter is described elsewhere. Weaver came back to the University from Throop (California Institute of Technology) in 1920, as an assistant professor, and succeeded Van Vleck as chairman of the Department of Mathematics in 1928, while Slichter was still dean. From 1932 until his retirement, he was with the Rockefeller Foundation and became its vice-president.

After Slichter's death, Weaver wrote to Mrs. Slichter: "I think you know that he has always been one of my major heroes. I have always felt that I owed him more than any other teacher and counselor and friend." And in writing to Louis Slichter he said, "I have had a chance in my position to know a good many men, and I have had a chance to know a great many interesting men; but I have never known more than one or two or three really great men. Your father was in that list, as far as I am concerned, even though I narrow it down to a single person."

Karl Paul Link was born in 1901 and took Agricultural Mathematics in 1918–19. He particularly remembers Slichter approving his correctly answering a question from his knowledge of plowing rather than of mathematics. Link said that if he could sum up Slichter's philosophy of teaching in one sentence it would be, "I'm going to try to make mathematics easy for you, and interesting." (In connection with Perry Wilson you will see Slichter's own interpretation of this same subject.) Link received his Ph.D. in 1925 under Professor W. E. Tottingham,

During his graduate work Slichter called him in several times to ask him how things were going and to draw from him comments about the faculty. On December 10, 1930, Slichter wrote to Thomas E. Brittingham, Jr., suggesting that the trustees of the Brittingham funds make an allocation to support the work of Link for five years. It is of course probable that Slichter had conversations on the subject with Brittingham before the letter was written. It is such a remarkable document in the details of its provisions and the breadth of understanding of biological research in the University, as well as in backing a brilliant young scientist, that it is quoted in full:

~ I take the liberty of suggesting (1) that from the Brittingham funds for which you are trustee, there be set aside for the next five years beginning July 1st, 1931 the sum of $27,500 for the support of a Brittingham Research Professorship in Bio-Chemistry, (2) that this professorship be allotted to Mr. Karl Paul Link, at present Associate Professor of Bio-Chemistry in the University of Wisconsin, (3) that a salary of $5000 for each of the first two years, $5500 for the third year, and $6000 for the fourth and fifth years of the tenure of this professorship be set up, (4) that it be definitely understood on the part of Mr. Link that the program is to be continued for the complete period of five years and not interrupted by resignation or otherwise, (5) that it be definitely understood on the part of the University that their present contribution to Mr. Link's salary, amounting to about $3000 per year, be assigned to that department for an assistant or assistants to Mr. Link as he may desire.

In explanation of this suggestion I will state that there has been under way in various laboratories of the University of Wisconsin for a number of years a study of a group of fundamental problems in the Chemistry of Vital Processes in which results of great scientific importance have already been obtained and which we are very desirous of developing further in the future. The work of Mr. Link forms an important part in that program. In other words we are undertaking at the University to make a many sided approach to the complex phenomena of the living cell. Work in hormones under Professor Hisaw, in vitamins under Steenbock, and in viruses under Duggar seems to open up an entirely new field of work which I believe will become the dominant field in biological science for the next generation.

Forty years ago a revolution was created in the science of Physics and Chemistry by the discovery of the electron, one of the results of Roentgen's discovery of the X-rays. There resulted the enormous development of atomic and molecular physics which has created a domain larger than the old domain of the science of Physics. At the present time we are just making a beginning in the study of biological phenomena where the basal action is associated with elements that lie between the molecule of the physicist and the cell of the old biologist. In the opinion of many this field of biology will soon develop to occupy a domain larger than that of the old field of that science.

Many of the elements involved in these studies have to do with material that passes through the finest porcelain filters. Some of these are termed viruses because they seem to be associated with the destruction of the normal life of the living cell. Others are called bacteriophages and seem to act favorably in promoting normal cell life. Whether these particles are living or non-living is a mooted question. We do know that most minute quantities of them produce phenomenally great results. It is doubtful whether any of the viruses have yet been obtained in a pure form. It is also true that probably none of the vitamins have yet been obtained in a pure form. Perhaps more definite results have been obtained with the hormones. In any case, we are here dealing with agents that are minute in size and excessively minute in quantity and that control in a very mysterious way the phenomena of cell activity.

Mr. Link's work lies in the chemistry associated with living matter, especially with plant life. In our approach to the phenomena of the vital processes it is desirable to emphasize the work in plant material because plants are very cheaply grown, and large quantities of viruses and other material desired can be cheaply obtained and experimented upon under controlled conditions. It is not intended that Mr. Link should devote his attention to this work merely as a helper to others, but he is to be permitted to carry on his researches in any direction that opens up, which from time to time seems to indicate the most favorable line of advance. Mr. Link's work has been especially associated with what we may call plant chemistry. During the last year he has printed seven papers having to do with the phenomena of the production of organic compounds in the living plant. A strong community of interest in all the allied problems in the fields of both plant and animal physiology, bio-chemistry, physical chemistry, experimental zoology, and plant pathology already exists and is of long standing and has drawn the departments into active cooperation. It is obvious that this cooperation can not be forced but must take place in the natural course of events by the similarity of the ends sought and by the enthusiasms of the various research workers.

Mr. Link is a young man well trained abroad in organic and bio-chemistry. He has studied at St. Andrews in Scotland, in Switzerland and in Vienna. As far as it is possible to predict human affairs, we believe it safe to appoint him to this position of honor and to be reasonably certain that his five years of research will be both productive and of a character materially to advance this important field of science and that the results will be of satisfaction to the donors of the fund.

The Brittingham Trustees agreed to the proposal, although they cut the salary figures somewhat, and later added funds for additional equipment. Slichter continued to back Link's work.

Early in 1966, David F. Allmendinger interviewed Link and had a fascinating time. We have used some of his report above. Allmendinger reports the following episode: Link once had a sophomore student, in the depression years, who was truly a great chemistry scholar. Link had hired him at twenty-five cents an hour to work in the lab, and noticed the boy was doing his own research on the side—and doing well at it. So, Link went to the dean and asked if he could give the student, although only an undergraduate, an appointment normally of the type granted to graduate students. The dean told him to send the boy up to the office and he would see. After fifteen minutes Slichter called back and said, "You can do it, but if he doesn't work out I'll fire both of you!" Carl Niemann was not fired, but instead became a member of the National Academy.

Link credits Slichter with being the chief builder of the Graduate School, but, of course, with the aid of a distinguished faculty.

Perry Wilson was one year younger than Link. He received his doctorate under Professor Edwin B. Fred in 1932. It was natural that Fred, who was both a neighbor of Slichter's and a member of the Research Committee (he was to become Slichter's successor on the way to the presidency of the University), should tell Slichter about his brilliant student. A good account never lost anything by passing through Slichter, and on May 2, 1928, he wrote President Frank, "Allot $1,000 to Professor Fred for the services of Perry W. Wilson. Professor Fred has discovered in Mr. Wilson a scientific genius perhaps more remarkable than anyone that has been at the University in a great many years. He desires to keep him on productive research at Wisconsin rather than permitting him to go to another institution." Years later when Wilson published a book on nitrogen fixation Slichter wrote to him:

~ My dear Wilson: Dr. Fred brought your new book, "The Biochemistry of Symbiotic Nitrogen Fixation" to me before Christmas. I thought that I would look over your method of organization, but I began to read and found the book so interesting that I have already read two-thirds of it. Your delightful style is both a scientific and an English model. You make the matters so interesting that I feel I must read the book and I even feel that I understand it!

You must be a fine teacher. Any subject is easy if made interesting and any subject is difficult if a teacher bores the student to death. I congratulate you on one of Wisconsin's most noteworthy books.

The hold that Slichter had on imaginative young men is illustrated by Wilson's reply:

~ It was very thoughtful of you to write me regarding my recently published book. I was highly complimented and pleased to learn that it met with your approval. I value your kind remarks for a personal reason. Several years ago when I was still an undergraduate, you talked informally to a group in our bacteriological seminar on the subject of Science and Authority. I was so excited with your delightful lecture that I persuaded Dr. Fred to secure me a copy of the manuscript. On reading it, I was once more so struck by the clarity and charm of the exposition that I resolved that I would always keep ahead of me the goal of presenting scientific material in a clear and stimulating manner. To have progressed toward this goal sufficiently to merit your praise is assuredly a great source of pleasure.

I have known for a long time that Slichter was one of the heroes of Homer Adkins, the great organic chemist of the University of Wisconsin, but I did not know until recently that, in January 1934, the Dean had written to Adkins, "I can assure you that I am probably even more proud of your work than you are yourself. It is a great laboratory and it is also a great university you are serving."

It was not only natural scientists whom he encouraged. Many academicians scorned the field of education. Not so Slichter, who wrote to a superintendent of public schools in Arizona: "Education itself is becoming one of our most important sciences and one of the important fields of research. It should be adequately supported at the Universities, and those entering the fields of Public Education and College Education should have adequate training in this field." As mentioned in Chapter 3, he did not believe that courses in education should be required for the doctor's degree. But when he found an able young man such as John Guy Fowlkes in the Education Department (later, Dean of the School of Education), he comforted and abetted him and gave him warm human understanding in the struggle against mediocrity.

When I became Dean of the College of Letters and Science at the

University of Wisconsin in 1942, I received from Slichter a note too nice for me to quote, but it elicited from me a reply which, in speaking of my zest in entering upon the new work, said, "Some of it comes from a feeling that Wisconsin is really a vital university. Vital intellectually, and vital in its care for personality and the joyous things in life. That feeling is the resultant of many forces, some of which are totally unfelt or at least unrealized by me, but some few stand out. Your own work, but much more your own self are [sic] one of these that make me care about the University you love."

There were other men somewhat nearer his own age whom Slichter influenced decisively—none of these more than Max Mason, who was thirteen years his junior. Mason, whose work on submarine detection we have already described, had been one of Slichter's students and was the major professor of Warren Weaver and of Slichter's son Louis. At every step, from student to President of the Rockefeller Foundation, Mason was cheered and aided by Slichter. Much as Slichter wanted him to stay at Wisconsin, he backed him for the presidency of the University of Chicago. In a letter written to Dean Henry G. Gale of Chicago on July 14, 1925, Slichter ascribed to Mason almost every human virtue. In describing Mason's unselfishness, he reveals himself. "This virtue I rate very high indeed in the qualities desired in a university leader— for in our evolving universities a great leader is called upon to sacrifice himself absolutely—to use himself as fuel in that great engine of humanity." After mentioning Mason's charm, he continues:

~ Next I must mention Mason's versatility. I have always said Mason approached true genius nearer than anyone I ever knew. The quickness with which he turns to anything and masters it is uncanny. There seems to be no limit to this power of mastering things, whether it be unessentials (or are they essentials?) like golf or billiards, or a problem of science or of the university. Among us he is the outstanding man in helpfulness to students. He is the outstanding man of ideas in the improvement of university work. He is the outstanding man in the promotion of better life among the students. He is the favorite speaker before alumni gatherings. He addresses more meetings of this sort than anyone else. He is the most helpful man in the general stimulation of research throughout our staff.

Slichter predicts a rosy future for the University of Chicago, if Mason becomes its president. He nearly hinted that Mason could cast out evil spirits, but did not attribute to him the ability to turn water into wine or to walk upon it. His almost extravagant praise at least did no harm, for Mason was soon appointed President of the University of Chicago.

Michael F. Guyer was ten years younger than Slichter and a Wood-
ward Grove neighbor. Slichter strongly backed his promising work on
inheritance. On March 13, 1923, he wrote to Guyer:

~ . . . *It is my personal opinion that there is nothing going on in
the University that is as important and should mean as much to us
for the future. It seems to open up unlimited fields for bringing
about changes in organisms that may be permanently inherited.*

*I hope that you will permit nothing of routine character to interfere
with the prosecution of your work without at least laying the matter
frankly before the university authorities. You have but one life to live,
and these next ten or fifteen years should be made the most productive. I know of nothing of greater importance in our scientific lines of
work.*

This now appears to be undue enthusiasm, since in the end the investigation did not pan out. Without risk, enthusiasm is pallid.

The University of Wisconsin is justly proud of its influence on the
development of colloid chemistry in America. The now long-established
National Colloid Symposium started at the University under the initiative of Professor J. Howard Mathews, for many years chairman of its
department of chemistry. The development of ultracentrifuge work in
this country began with a visit of Professor The Svedberg of Upsala,
who was also a participant in the first symposium. At the fortieth symposium Professor Mathews described the origin of the first:

~ *In 1920, a year after I became Chairman of our Chemistry Department, Professor C. S. Slichter was made Dean of the Graduate
School. As he was a dynamic person he naturally considered what
might be done to improve graduate work and to stimulate research at
Wisconsin.*

*One day early in 1922 I was in his office on some long forgotten
errand, when he said, "Mathews, we ought to do something to pep
up research at Wisconsin. We ought to have some symposia on subjects of interest—something different. You are interested in colloid
chemistry, how about a symposium on colloid chemistry?" Well, have
you ever seen a hungry trout go for a grasshopper?*

This conversation also led to Svedberg's visit.

Perhaps I have given too many illustrations of Slichter's relationships
with individuals, but it is these relationships which formed his greatest
contribution to the University. As Slichter grew older his loud voice
lost none of its volume or resonance; his deafness which had plagued

him for many years increased, yet he remained, perhaps, the most intuitively responsive listener on the campus.

Rules and Their Enforcement

Rules did not entrance Slichter, and credits were not his fetish. In reply to an inquiry from a professor at Washington University he said, "We take no interest in the division that a graduate student makes of his time. If he should happen to be devoting twenty hours a day to a single seminar, it will heartily meet the requirements of the Graduate Office and give the candidate full residence credit towards a higher degree."

In 1943 Helen C. White, by that time a distinguished Professor of English at the University, wrote to Slichter:

~ I suspect I have told this story to half the town, but I am quite sure I never dared tell it to you. I think that I probably should not have got my doctor's degree at all if it hadn't been that years ago, when I was an instructor and I asked you about how long it would take me to get my degree, you discovered that I had been working all summer on what afterwards became my dissertation and that I was still working away at it; and, though it was November, you sent me down to pay my fees and register for a semester's credit. I remember that that quite unexpected generosity made the Ph.D. seem feasible, and I have always been very grateful to you.

Warren Weaver in his autobiography recalls:

~ Following his most active teaching period, Slichter was an imaginative, if admittedly unorthodox, dean of the graduate school. In his office one day, he called me to his desk and said with evident glee, "Here's a chap I am certainly going to let into the graduate school. Look at his transcript! See what judgment he used in picking courses to flunk in!" I get a particular charge out of this story as I look at my eighth-grade report card, which lists 98 in drawing, 97 in algebra, English, and music—but a very low mark in both spelling and deportment.

Even in fiscal affairs he retained much discretion. In a letter to the National Tuberculosis Association he described three of his methods for handling funds entrusted to the Research Committee: (1) Deposit with the business manager—this leads to much red tape. (2) Ask the secretary of the regents to open a special account merely to assure donors that the accounting is businesslike. (3) Hand over the money to the investigator. This third method was used in the case of William Snow Miller, distinguished anatomist.

But when rules had to be enforced he did so with vigor. Upon the suggestion of his nephew L. C. Burke, he wrote to delinquents:

~ I have again received a notice from Mr. L. C. Burke, Ass't Librarian, that you have not yet paid your library fees for the current year. Your class cards are being held in the Graduate Office until we receive a notice that you have paid these fees. Unless the fees are paid within the current week, you will be obliged to forfeit credit in your graduate courses. This notice is final.

This was written on a Wednesday so that prompt action was required and usually received.

Faculty members whose grades came in late were particularly difficult. On May 23, 1928, he wrote to Professor C. E. Mendenhall of the Department of Physics:

~ I am very much embarrassed at the necessity of sending you about 50 report cards covering graduate work of students majoring with you in Physics, extending over the past two or three years. In many of these cases this represents the second or third call we have made upon your department for a report. We do not know in these cases whether the students have completed their registration and are deserving of a degree or not and cannot know until we have reports from your office. It is embarrassing to say so, but I must report that you are the only individual having graduate students that habitually puts the Graduate Office to this trouble. I am trying to keep down the cost of clerical work. This adds to it. You have clerical workers in your office whose duty it is to see that all the reports of instructors are properly forwarded at the close of each semester. The expense of looking after these matters amounts to about $25 and I am very much afraid I must find a method approved by the President of charging this expense to your department. I will greatly regret to take it out of your research fund but I . . . am unwilling to do in this office the work that pertains directly to the departments and for which university provision has been made.

When the grades of Professor John R. Commons did not come in on time Slichter went after his secretary, Hazel Farkasch,

~ I am sending to you a number of report cards calling for reports of graduate students now long past due. This work puts considerable expense upon this office which we are unwilling to bear. Especially it points to some marked inefficiency in the clerical workings of your office. It is the duty of the clerical staff of Professor Commons to see that all reports are made each semester. I regret to take the matter up in this way, but it will be necessary to charge back to your office the needless expense put upon me in this way, as I have no funds to take care of it.

The reply of Miss Farkasch was far from obsequious. She stated that she had assumed—I think quite properly—that Commons's assistant should have seen to the grades, but she promised compliance. Neither secretary nor dean thought it worth while to suggest that Professor Commons might attend to the matter, for he was notoriously casual about such details.

This was not the only use which the Dean made of a technique of threatening to charge for wasting of his secretaries' time. Questionnaires sometimes received similar treatment. For instance in 1928 he let it be known that the estimated cost of answering three of the questions from an out-of-state teachers' college was eighty-one dollars. The inquirer wanted to know how many earned Ph.D.'s had been granted by the institution, how many of the recipients were still living, what their names were, which of them were in *Who's Who*, and "What method of selecting Ph.D. 'timber' do you use?" In 1931 he reported to the Association of American Colleges that "the number of questionnaires coming to us has reached the vanishing point since we have informed the applicants that the cost of answering the questionnaire must be met by those applying for the material."

Relations with Colleges of the State

The University of Wisconsin received many inquiries as to the colleges from which it accepted graduates for admission to the Graduate School. Slichter's influence was always in favor of judging the man rather than the school from which he came. There were some colleges, however, whose graduates were scrutinized with particular care. He believed that some small impecunious schools had intelligent and devoted faculties along with hard-working students. He often proclaimed the virtues of the best of the products of Milton, of Northland, and of other worthy but, at that time, not fully accredited schools.

On July 15, 1925, in reply to an inquiry from the Department of Education of the State of Minnesota in regard to Northland College, Slichter stated:

~ . . . *this institution exists under very difficult conditions.*

There is no question that it is completely imbedded in poverty. The students in the institution have no money, almost without exception they work their way through school in large part by working on a farm attached to the college. Notwithstanding the fact that this institution could hardly be rated high on the basis of its physical plant or endowment, I am always glad to have one of their graduates present himself for admission to the Graduate School. The members of the faculty are inspired by a spirit of the old-fashioned college and take an almost fanatical interest in their work with the students.

I have taken an interest in this institution because of the spirit of hard work and thoroughness with which everything is done in spite of the numerous obstacles. For the purpose of admission to the Graduate School I rate their graduates very high and I believe you would be entirely justified in receiving applications for certificates from the graduates of that institution on the same basis as those received from Beloit, Lawrence and Downer. In fact, I believe in many respects you would find their graduates superior.

And on May 4, 1928, Slichter wrote to President Whitford of Milton College:

~ I *understand you desire a statement from me concerning the work of Milton College and its reputation for scholarly work. I am very glad indeed to reply to your request, because I have known of the work of Milton College for many years. I was well acquainted with the staff that preceded your time and know a great many of the present faculty personally. There is no college in the west concerning which I have a higher opinion. Your institution has always done thorough work and has done much to keep alive the moral and intellectual ambitions of the young people that have come to you for guidance.*

Since 1920 as Dean of the Graduate School I have been able to judge of the work of your college from the large number of your graduates who have registered in the Graduate School at the University of Wisconsin. Since that time we have had 114 registrations from your college alone, representing about 60 individuals. A number of these have proceeded to the degree Doctor of Philosophy and we hold these among our most prized graduates. During all this time we have not had a single graduate of Milton College who did not do good work in our Graduate School, a statement which I can hardly make concerning any other college. You have known from me personally of the high opinion I hold concerning your guiding influence in the work of Milton College and it is not necessary for me to en-

large upon it here. *The colleges of the state have a distinct field of work highly important to their communities and should receive the support of everybody who is interested in advancing the best things among the young people of the state.*

His enthusiasm for good, small colleges led him to try to interest people in them. In Chapter 10, in connection with his charities, his gifts of books to the University of the South, Northland College, and Washington and Lee are described.

Carl Schurz Professorship

In 1911 a group of Americans of German descent, many of them from Milwaukee, gave to the University of Wisconsin a sum of about $30,000, later augmented by further gifts and the additions of some of the income to capital, to establish a chair known as the "Carl Schurz Memorial Professorship," named for Carl Schurz, a German liberal who had come to America and for some time had lived in Wisconsin. He had been a Regent of the University. The chief use of the income was to provide for bringing, from time to time, scholars from German universities to the University of Wisconsin as visiting professors.

There were, of course, no such visitors during the participation of the United States in World War I.

By the fall of 1921 the question of resuming such appointments was being raised. Professor A. R. Hohlfeld, Chairman of the Department of German and one of the University's most distinguished faculty members, wished to do so; moreover, he was to spend the spring of 1922 in Germany where he could negotiate with possible appointees. It was first proposed to bring a biologist, but Professor C. E. Allen of Botany did not wish any German visitor so soon after the war. Early resumption was espoused not only by Hohlfeld, but also by Slichter, who took a leading part in pushing the cause and selecting the professor. On December 14, 1921, he wrote to Professor L. R. Jones of plant pathology: "Professor Allen's objection is entirely general. He objects to a German coming here in Mathematics as well as to one in Botany irrespective of previous history or personality or eminence of the scientist. As his objection seems to be entirely temperamental, I see no reason why we should stand off from making the exchange professorship if a suitable candidate can be found outside of the immediate field of Professor Allen's work."

The physicists were anxious to have such a visitor. Soon it was decided that the first choice among those available was Professor Arnold

Sommerfeld, at the time Professor of Physics at Munich. Birge clearly supported Hohlfeld and Slichter, and secured the Regents' approval of this course of action. Professor Sommerfeld thus became the first postwar Carl Schurz Professor. It was a most fortunate appointment.

The willingness of Slichter to renew normal relations with the Germans was well in advance of the attitude of many others in Wisconsin, and is noteworthy in light of his war activities in connection with the work on submarine detection. His attitude may in part have been derived from his psychologically satisfying ability to focus his animus on some individual—in this case, the Kaiser, whom he classified as just below the devil, but less of a gentleman. However, with Friedrich Wilhelm Viktor Albert, by the Grace of God, King of Prussia; LL.D. University of Pennsylvania, safely chopping wood in Holland, the normal relationship between peoples should be reestablished. Moreover, even during the war Slichter had published a pamphlet with Professor George Wagner on *Germany's Gain from Germany's Defeat*, which showed his hatred of the German political system but his warm regard for the German people.

Association of American Universities

The Dean took an active but not a leading part in the affairs of the Association of American Universities. His idea for a program committee was given to Dean A. H. Lloyd of Michigan, "I would prefer that committee to consist of yourself and your own shadow."

Perhaps his most notable contribution to the AAU was a presentation on November 10, 1927, to their 29th Conference, of a short paper entitled "Debunking the Master's Degree," the main purpose of which was to protest against down-grading the education of teachers. The paper starts out with:

~ *The college and university catalogues of fifty years ago contained a statement that read about as follows: "The degree of Master of Arts will be conferred on Bachelors of three years' standing who have sustained a good character, and furnish satisfactory proof of having pursued professional or other advanced studies. Application should in all cases be made to the President before Commencement, accompanied by a fee of five dollars." The good character required meant, in practice, that the candidate had successfully kept out of jail, and the pursuit of professional or other advanced studies was taken as a matter of course. The real requirements were the lapse of time and the payment of five dollars. There is no record of a fee of five dollars*

*"received by the President before Commencement" being returned to
the candidate with a refusal of a Master's diploma. This was in the
Victorian era. Then at the close of that era—in fact, during the "gay
nineties"—came a revolution in the whole field of higher degrees. The
Master's degree was rather abruptly changed to a degree "in course"
instead of a degree "of course." A year of resident study and a thesis
or essay were quite generally set up as the new requirements. These
new standards were primarily the result of the growth of graduate
study at Johns Hopkins University.*

This last sentence brought a polite but firm protest from A. Lawrence
Lowell, President of Harvard, who wished to have that institution either
share or take the credit given to Johns Hopkins.

When Slichter described how to get the master's degree before the
"gay nineties" he knew whereof he spoke, for he received a master of
science degree from Northwestern University in 1888, encouraged by
the catalogue, which contained, in essentially the same words, the
statement he quoted above.

"Debunking" then describes how various schools were giving master's degrees to schoolteachers on a basis of summer school work, often
a rather small amount of it, and for correspondence study work. He
pointed out that the master's degree in engineering and in other technical subjects had been kept at the high level, but this was not done for
teachers. To this he objected.

~ *To defend and protect the teaching profession we must not only
test out the intellectual qualities of the candidates, but must also see
to their technical equipment and be able to appraise their personal
and spiritual qualities. . . .*

*We cannot take too seriously the importance of the profession of
teaching, if we are to take seriously the future of the American experiment in democracy. It is not alone the personnel in the score or more
large universities that counts. More important to the welfare of the
state is the personnel in the thousands of secondary schools of this
country—the teachers whose duty to youth is "to declare the things
of life as the dawn first glows with the coming day." It is vital that we
lead to the staffs of our public and private schools of all grades the
highest type of men and women—persons of ambition, scholarship,
and sterling, virile character—and that we continually shunt away
from this profession the weaklings we cannot trust with great duties.
A part—perhaps a large part—of this responsibility leads back eventually to the institutions of higher learning. The standards and attain-*

ments *required for the degree primarily sought by teachers are not
a matter of indifference.*

In light of how he had received his own master's degree, Slichter
should have been sympathetic with the following request which came
on pink paper from the "Palace, Maihar State" in Central India, dated
April 6, 1926.

~ *Sir,*
   Will you kindly grant me the degree—M.Sc.?
   In return I shall make a contribution to your University.
   Hoping to be complied with

                                        I remain, Sir
                                        Yours obediently. . . .

We have no record of a reply.

A Presidency Declined

In 1927, R. E. Tally, a mining executive and chairman of the Board
of Trustees of the University of Arizona, approached the dean as to the
possibility of his becoming President of that university. It was an attractive prospect. Slichter was fascinated by the problems of arid regions
and the spirit of the Southwest appealed to him. Max Mason, who had
worked with Mr. Tally on some geophysical surveying, had suggested
Slichter's name. Since both the chairman of the board and the chairman of the faculty search committee wished him to be appointed, it is
almost certain that, had he indicated a willingness to accept, he would
have been offered the position. He inquired through Philip F. La Follette concerning the politics of Arizona and received a discouraging
description of its intricacies. He visited the campus. But, although he
was tempted, he finally declined to allow his name to be presented.

His son Louis believes that Mrs. Slichter was more certain than her
husband that they had roots too deep to survive transplanting. As so
often, she was both wise and persuasive.

Pride in the University of Wisconsin

Both when serious and when playful—he was often both at once—
Slichter took great pride in the graduate work at the University of Wisconsin.

Although he defended the College of Letters and Science from losing funds to the College of Agriculture in the days when Governor

Hoard was on the Board of Regents, he recognized the magnificent accomplishments of the latter college. He wrote in 1925:

~ . . . The two institutions in America that stand out as graduate institutions in Agriculture are Cornell and the University of Wisconsin. I can say that the number of graduate students in Agriculture in the University of Wisconsin is greater than that in any other institution and that the work in the following departments is unequaled and unexcelled in any institution in America: Plant Pathology, Biochemistry, Agricultural Bacteriology, Agricultural Economics, Nutrition, Genetics, Animal Husbandry. The work in the departments of Soils and the various fields of Dairying are also very strong.

In 1930 he said in an address to the Sigma Xi Chapter at the University of Wisconsin: "The University's mission is to develop a minority which shall be critical and skeptical, eager and inquiring, and acutely conscious of ideas."

A year before his retirement, Slichter, in a more jocose mood, replied to an inquiry from Northwestern University:

~ My report must be the least sensational of all, as I have been continuously at college and in the classroom since my freshman year; and I have had, after Evanston, but one residence. In consequence, I think my classmates would find me greatly improved—except in morals and penmanship—otherwise college life must confess itself a failure.

My four sons are all men now, so in order to seek my mental level, I must fall back to my grandchildren for companionship.

I am very proud of my Graduate School, which I can easily prove is the best and biggest anywhere, for my stock of modesty has greatly grown since I forsook the intimacies of the great of Eighty Five.

Defense of Research Budget

A university administrator is constantly defending his budget against those who would despoil it either by taking it for, or making it assume, the obligations of other units. It also must be defended from expenditures which are either inappropriate or not of sufficiently high priority.

When the depression hit, each dean tried to persuade President Frank that his budget should not be decreased substantially. Thus on December 7, 1932, Slichter, who had recently survived the ban on foundation gifts, wrote to the president.

~ I have just learned that the University Research Fund has been reduced in the legislative requests from $75,000 to $50,000. You will recall that I especially requested that this fund be not reduced without consultation with the University Research Committee. It is very unfortunate that they have not had an opportunity to be heard in the matter, as this is the only fund whose administration is in the hands of a University committee, and this committee has devoted itself diligently to the economical distribution of the fund. In fact, the governor of the state [Philip La Follette], who has devoted unusual attention to the study of the University budget, has stated that this is the best administered fund in the University. It is also the fund whose reduction will do most to destroy morale among the ablest members of the University staff. I trust you can find some way to avert such misdirected economy.

It had been a hard struggle to get "Number 9" up to its then current level. For instance, in 1926 Slichter had written to the president: "As you know, the fund at present available for allotment by the University Research Committee amounts to $30,000 annually. This sum has proved entirely inadequate to meet the needs. In fact the demand from projects that should be supported has become so great that it has become very embarrassing indeed to allot the money."

In 1931, Sellery, for whose position I have hereditary sympathy, suggested that the Research Committee take over Hisaw's salary. Slichter's comment was:

~ After the University Research Committee had completed its labors, I received the copy of a letter from Dean Sellery suggesting that the University Research Committee place Mr. Hisaw's salary in the research budget for the coming year. I immediately called the committee together. They are of the unanimous opinion that such an arrangement is not desirable. We should avoid taking over any salary on a permanent or semi-permanent basis in the University Research budget. The committee interprets the University research fund as a means of support to projects immediately in need of aid and where adequate aid is not available from any other source.

Slichter's own prescription for saving money was to curtail the extension activities of the University, doing away with its dean, and leaving its administration to the colleges, with a coordinating committee of the faculty. Also, as early as 1925, he pointed out that the proliferation of courses was an evil, but one that he did not know how to control.

In those days, even before the depression, when research money was scarce, constant vigilance was required to keep certain costs down, particularly expenditures for printing and travel. For example, in 1923 Slichter had refused to E. B. Fred an allotment to buy reprints, a refusal that brought a strong but unsuccessful protest.

At a time when institutions were turning to industry to support research, Slichter strongly disapproved such arrangements with individual corporations but favored plans for securing funds from industrial associations or other groups.

Refusal of Foundation Gifts

On August 5, 1925, the Regents of the University, after a long debate and against the united wishes of the University administration, passed the following resolution: "RESOLVED: That no gifts, donations, or subsidies shall in the future be accepted by or on behalf of the University of Wisconsin from any incorporated Educational endowments or organizations of like character." The Regents passed the resolution by a vote of nine to six.

The immediate effect was to make it impossible to accept a grant from the Rockefeller Foundation for the work of Professor Arthur Loevenhart on syphilis. How many other grants would have been secured during the period until March 5, 1930, when the resolution was rescinded, can never be known though it is believed that a large amount for medical work was forfeited by the action. Moreover, no one can assess what the loss of momentum in securing funds cost the University, though the development of the Wisconsin Alumni Research Foundation and the quality of the faculty prevented irreparable long-time injury in this regard.

Slichter was active both in the unsuccessful attempt to block the unfortunate action and in getting it rescinded, as well as in minimizing its current and its lasting effects. While it was in effect he kept needling the president, thus in explanation of the item for the work of Professor Loevenhart in the Graduate School budget for 1927–28 he wrote, "This work has been materially set back by failure to accept $17,000 a year from the Rockefeller Foundation. The project should be prosecuted vigorously and cleaned up completely in the next two or three years. There is nothing gained by inadequate support. The quicker the University can profit from this work the better." As soon as grants could again be accepted, he was stimulating and supporting major projects.

The self-denying ordinance of the Regents, followed by the depression, precluded Slichter's raising substantial amounts of money for research from sources other than the State, or, as detailed later, the Wisconsin Alumni Research Foundation. However, within a week of the rescinding of the Regents' action, he was planning a drive to support research in hormones, vitamins, and filterable viruses.

Wisconsin Alumni Research Foundation

Establishing the Foundation. The Wisconsin Alumni Research Foundation (WARF) is a nonprofit corporation established in 1925 to manage patents given to it, and the funds derived from them, in the interest of research at the University of Wisconsin. The original patents were given to the Foundation by Professor Harry Steenbock. Since that time, WARF has granted to the University over fifty-five million dollars. In 1970 these grants amounted to over three million dollars.

It would be easy to write a controversial account of the establishment of the Foundation. A number of persons played essential roles at the start. Friends of the protaganists have emphasized their roles. Steenbock, in desiring that Slichter should get full credit for his part, made statements that were not in complete accord with his later memory. The fact that human memory is fallible accounts for some of the difficulty in unraveling the strands, but perhaps a greater difficulty lies in the impossibility of knowing how an idea develops when people are working together. After discussing problems, several persons may independently see what the next step should be. This is particularly true of people as able as Steenbock, Russell, and Slichter, and such men as Harry Butler, Louis Hanks (official name Lucien, but he preferred Louis), George Haight, William S. Kies, and Gerhard Dahl, whose enthusiastic collaboration they soon secured.

It is my opinion that Steenbock first, and Russell next, and perhaps more clearly, saw the desirability of patenting Steenbock's discovery; that Steenbock and Slichter developed the idea of a foundation and heartily agreed that its board be composed of alumni and not include either faculty or University officialdom; and that it was Slichter, far more than the other two, who worked out with Harry Butler the articles of incorporation and the bylaws of the Foundation and selected the members of the board. Several had been among those who at Slichter's request had supported the work of Max Mason on submarine detection. It was Slichter, not without some reluctance on the part of the other two, who channeled the expenditures into the hands of the Research Committee of which he was Chairman. I have made no attempt to marshall all the evidence, some of it hearsay, for or against these conclusions, but have used some to tell the story of what happened.

Steenbock, Russell, and Slichter were able and strong-willed men and they worked together with remarkably little friction. No one of the three was deterred by the inevitable criticism of patents in the field of public health. Slichter had to an unusual degree one quality that was not as evident in the other two—the ability to become enthusiastic about another man's ideas, to make this enthusiasm contagious, and to carry these ideas beyond their logical conclusion. This quality was essential to the whole enterprise.

It is perhaps well at this point to sketch the career of Steenbock prior to 1925 and to mention a few of his characteristics. He was born in 1886 on a farm in Calumet County, Wisconsin, a pleasant land of rolling hills sloping down to the east shore of Lake Winnebago. He entered the University of Wisconsin in 1904, took his bachelor's degree in 1908, and his master's in 1910. He received the doctor's degree, also from the University of Wisconsin, in 1916. Among the thirty-four receiving their doctorates that year, a number later had long service with the University, and two besides Steenbock—Vernor Finch of Geography and Henry Schuette of Chemistry—also served on the University faculty for over forty years. In the period before he received his doctor's degree, Steenbock studied at both Yale and Berlin, and was a full-time instructor at the University of Wisconsin starting in 1913. He was an Assistant Professor in 1916–17, Associate Professor from 1917 to 1919, and Professor or Professor Emeritus from 1919 until his death in 1967. The title of his dissertation, "Studies in Nutrition," can serve as a title for his lifework. In the faculty memorial it was said that he "had the vision to patent his irradiation discovery as a means for preventing its abuse and exploitation, and he made certain that the proceeds from the patent be invested to yield income which would foster additional research." The memorial also said:

~ It is difficult to describe the personality of Steenbock for it had many facets. He could be rigid in dispute and gracious in companionship. He had high standards to which he adhered and which he was not unwilling to press on others. He believed intensely in the freedom and responsibility of the individual: for instance, he would explain how with him it was a matter of conscience to devote his discoveries to the benefit of mankind and the income derived from them to the University, again for human betterment—but that he believed no one else had either the legal right to force him to do this or the moral right to expect it of him.

He insisted that the Trustees of Wisconsin Alumni Research Foundation have full authority and that the scientists who benefited from grants be accorded the greatest degree of freedom. When, in his judgment, errors were sometimes committed or liberty abused, his criticism

*could be harsh; but it never made him forsake the principles of indi-
vidual freedom. Although his point of view was often conservative in
the extreme, he had the thrust of the future in his actions.*

*A tall, erect figure, a mind far beyond the ordinary, a straightforward
dominating personality; he was passionately devoted to the welfare of
the University of Wisconsin.*

It is easy to compile a paradoxical list of adjectives that fit Steenbock's nature—loyal, irascible, gracious, abrasive, rigid, sensitive, and
generous. He could hate, but he had deep and lasting affection for his
friends, and Slichter became one of them.

I give a mere outline of events connected with organizing the Wisconsin Alumni Research Foundation. Where Slichter is concerned, I
go into greater detail. This is not bias, but focus.

In 1923 Steenbock found evidence that irradiation by ultraviolet light
increased the antirachitic properties of various foods, including some
fats—in particular, that bane of mothers, cod-liver oil. The effect of
cod-liver oil and of irradiation in preventing rickets was known, but the
fact that irradiation worked through increasing the antirachitic properties of food had not been realized. The importance of this discovery,
if validated, was clear to Steenbock and to Dean Russell, whom he consulted. Its value was also clear to the Quaker Oats Company, whose
chief product could be fortified by this means, and they offered him a
contract that might have netted him nearly a million dollars for the exclusive rights to the patent.

Some years before, in connection with another discovery, Steenbock
had unsuccessfully approached the administration of the University,
suggesting that they aid him in securing a patent. He spoke to the president about patenting the new discovery but received no encouragement.
By this time he was aware of the fund that had been established at
Toronto by Banting at the time of a development of insulin, and of the
Research Corporation of New York organized to handle patents for
universities and their faculties. Steenbock's discussions with Carl S.
Miner, of Chicago, and others also made him believe that such organizations were worth considering. Moreover, the important problem of
controlling a patent in the interest of the public was difficult. The lack
of such control in the case of the Babcock milk test had, in his judgment, not been in the public interest, but the exercise of such control,
unless the value of the patent were to be diminished by the granting of
an exclusive license, would not be within the ability of a single person
nor appropriate to the duties of a professor. The defense of a patent
against legal attacks or infringement would again be beyond and out-

side the scope of Steenbock's activities. Yet Steenbock greatly desired that the gains which promised to be derived from the patent redound to the benefit of scientific research at the University. The cost of securing patents, both in this country and abroad, was beyond what Steenbock felt he should afford. However, he decided to apply for the United States patent, and the first application was filed June 30, 1924. Later the application had to be amended before the patent was granted.

In summary, Steenbock wished to secure funds with which to pursue the application for the patents, both in the United States and elsewhere; and to find an organization which would protect the public from fraudulent or ignorant use of his discovery, and would manage and defend the patents and exploit their financial possibilities in the interest of scientific research at the University. However, it had become clear to him that the University was unwilling, and probably unsuited, to act as agent in this regard.

At what point Steenbock first talked with Russell about this matter is not clear, but probably he had been in touch with him at the time of the earlier attempt to interest the administration in the patent question. Certainly Russell was kept well aware of the whole problem as it unfolded; and he discussed the situation with Slichter, who was not unacquainted with this work, for he had studied applications for support of its early stages. A grant had been made to Professor E. B. Hart and Steenbock, in January 1924, for histological studies in connection with rickets. In an undated application forwarded by Slichter to President Birge in January 1924, Steenbock asked for $600 for a half-time assistant to help in the study of the effect of light on bone structure. This was when the truth, as so often accompanied by fable, was emerging, and Steenbock wrote: ". . . It has been found possible to confer growth promoting properties upon a ration of purified food constituents by the simple expedient of exposure of the ration itself to ultra violet light. Not only that but certain properties have been conferred upon galvanized wire screens as a result of which these screens stimulate growth." The grant was approved by the Research Committee.

In the spring of 1925 Slichter called Steenbock and invited him to lunch at the Madison Club. The story of their meeting is perhaps best told by Robert Taylor, now a vice president of the University, who had a series of interviews with Steenbock in 1956:

~ *Slichter was waiting at the Madison Club, when Steenbock arrived.*

"Tell me all about it, from beginning to end," the dean started. And Steenbock did, omitting no details on how the University had refused

to provide money for the original patent application, and how he proposed to set up a foundation to handle the patent, once it could be
used commercially.

"It's the most amazing thing I've ever heard," the dean commented,
"from now on, Harry, you worry no more about money."

"I'll go to New York City right away," Slichter promised, "and I'll
come back with plenty of funds to set up your Foundation."

Although I find no other evidence that Dean Slichter ever called
Professor Steenbock "Harry," this is basically a believable account, for
Slichter, when enthusiastic, encouraged without stint and both decided
and acted rapidly. Moreover, in the February 1921 issue of the *Wisconsin Engineer*, Slichter had indicated his faith in the role which science
would play in the economic welfare of the nation.

~ The great growth of national wealth in the past generation in
America was primarily due to the taking possession by the people of
our natural resources—possession of the soil, the forests, the coal, the
minerals, and water power. But the great spurt in the winning of this
kind of wealth has now come to an end. It is now clear that some
resources are being depleted at a rate that exceeds new discovery, so
that it is even probable that less national wealth of certain sorts will be
handed down to our children than we received from our fathers. It is
doubtful, for example, if the reclamation of land by irrigation, and
drainage, and clearing of forest, now materially exceeds the rate of soil
depletion. Certainly in one resource—timber—there is important depletion. It is evident therefore that if the curve of increase of national
wealth is to go on rising at the same rate as in the past, then enormous
increments of wealth must come from a new source—from scientific discovery. The betterments, the inventions, the discoveries—whether of
new substance, of new processes, of new things, of new scientific laws
—in the industrial or scientific laboratories constitute the new sources
of wealth for the present generation of Americans.

One cannot read this without recalling the splendid title of Vannevar
Bush's report to the President of the United States—*Science, the Endless Frontier*—written about a quarter of a century later.

Slichter quickly set to work, traveled to Chicago and New York to
see alumni, and soon had raised about $10,000 from those who later
formed the Wisconsin Alumni Research Foundation. He himself advanced $2,000 so the foreign patents might be applied for.

Two major tasks in addition to securing the patents remained before
the Wisconsin Alumni Research Foundation could be in operation.

The first was to be sure that the project had the approval of the Regents. Birge discussed this with the Regents on April 22, 1925, and they referred the matter to their Executive Committee, which invited Russell and Slichter to present the proposal to them. The minutes of the meeting, including the action of the Committee which was later approved by the Board, follows:

~ *Special Meeting of the Executive Committee*
of the Board of Regents
University of Wisconsin, May 8, 1925

Deans Slichter and Russell presented a plan with reference to the administration of the results of University Research, particularly as applied to those researches that are of a character and of such importance as to warrant patents being taken out on the same. Such patents may be taken:

1. In the interest of the public at large to prevent exploitation or monopoly, or

2. Assigned by members of the University staff for the direct benefit of the University.

The University recognizes that its organization is not well suited to attend to the details of patent procedure; to defend patents in litigation and conduct the necessary business of completing the commercial utilization of patents. Immediate need for the consideration of this matter lies in the fact that the results of experimental work of Doctor Harry Steenbock of the Department of Agricultural Chemistry of the College of Agriculture are such that patents should be secured at once to protect these discoveries. Doctor Steenbock has indicated his willingness to assign his interest in large measure to the University, on the understanding that the returns, if any, will be mainly made available for the prosecution of scientific research.

In providing a working plan by which this and similar matters can be handled, it is proposed to create an organization on a broad enough basis so as to embrace any other propositions of a similar nature that may arise in the future.

It is proposed to organize a non-profit-sharing corporation or trust, the necessary capital of which will be contributed by alumni and friends of the University, the management of such corporation to be placed in the hands of Trustees. This corporation will be empowered to receive assignments of patents and all rights in patent applications; to secure patents in this and foreign countries; to carry on further work with reference to the commercialization of the patent processes; to take all the necessary steps to defend and utilize these patents in a way that will best subserve the interests of this foundation. The subscribers to this fund will be repaid their advances from the avails arising from the sales of rights, royalties, and the like, after expenses incurred in the transaction of necessary business operations of the corporation are defrayed.

It is understood that the primary purpose of the corporation is to administer the patents intrusted to its jurisdiction in such a way that research in the University may be supported from the avails in excess of sums deemed necessary by the Trustees for the most beneficial development of the patents. It is understood that members of the University Staff who may assign patents secured in their name to this corporation for the ultimate benefit of the University research will be permitted to nominate the fields of research to which such funds shall be allotted.

The Executive Committee adopted the fundamental principles of the plan presented as affording a practicable means of the possible endowment of research through funds arising from the scientific efforts of its staff members who are willing to assign patents for such purposes.

Second, the actual charter of the Foundation had to be worked out. Five persons played a leading role: Slichter, Steenbock, Russell, George Haight of Chicago, a leading patent attorney, and the Madison lawyer, Harry Butler. Russell had had conferences with Butler and Slichter during the early stages, but was in the Orient when work was concluded. It would seem that the details were mostly hammered out by Slichter and Butler, who greatly enjoyed working together, and represent a document fully in accord with Slichter's views. The Articles of Organization were filed with the Register of Deeds of Dane County on November 14, 1925, by Thomas R. Brittingham, Jr., of the class of 1921, Lucien [Louis] M. Hanks of 1889, and Timothy Brown of 1911, with whom the Slichter boys had often sailed. In addition, the original trustees included George I. Haight and William S. Kies, both of the class of '99. It is almost certain that all of them had known Slichter since they were undergraduates, and it is likely that some of them had been his students. A Certification of Incorporation was issued by Fred R. Zimmerman, Secretary of State, on November 16.

The purpose of the Foundation was described as follows:

~ The business and purposes of the corporation shall be to promote, encourage and aid scientific investigation and research at the University of Wisconsin by the faculty, staff, alumni and students thereof, and those associated therewith, and to provide or assist in providing the means and machinery by which their scientific discoveries, inventions and processes may be developed, applied and patented, and the public and commercial uses thereof determined, and by which such utilization or disposition may be made of such discoveries, inventions and processes, and patent rights or interests therein, as may tend to stimulate and promote and provide funds for further scientific investigation and research within said University or colleges or departments thereof. . . .

Slichter was clear on the composition of the Board of Trustees. In 1928 he wrote to President Elliot of Purdue: "The one essential thing in the success of such a foundation is, in my opinion, to see to it that the board of trustees contains no member of the academic staff or administrative staff, including the president of the university, nor any members of the board of trustees of the institution."

Steenbock seems to have desired the inclusion of a statement concerning the protection of public interest and somewhat reluctantly agreed to the simpler form. He refused in the initial period to receive any income from the patents.

One element that Slichter hoped would grow within WARF, but did not to any great extent, was the receipt of annual gifts from alumni for the support of research. He had mentioned such a project to the chairmen of departments as early as March 27, 1922. This was one of those good ideas which the broom of history sweeps aside, but sometimes only under the rug. Increased support for research and the comparative growth of other University needs led to the establishment in 1945 of the University of Wisconsin Foundation, which incorporated the idea of continuous alumni support but has thus far directed its efforts chiefly, but not entirely, to enterprises other than research—such as a free fund supported by annual gifts and special projects, for example, the University of Wisconsin Center, the Elvejhem Art Center, and the Alumni House. Starting with the Frederick Jackson Turner Professorship, the University of Wisconsin Foundation has supported a number of professorships, some of them placing stress on research in the social studies or humanities.

In 1925 the lip was still far from the cup, for the patent was not granted until 1928, although Steenbock had formally assigned his right in it to the Foundation on February 18, 1927. Immediately thereafter he wrote Slichter: "Allow me most humbly to declare my unbounded appreciation of the generous material and spiritual support which you have given this project."

From this point onward the dominant figure in the business management of the patents was Harry Russell, who became Director of the

Foundation in 1930 and held the position until 1939; in investing the income from the patents to form an endowment for research, Thomas E. Brittingham, Jr.; and in spending the money, Dean Slichter. Each was remarkably fitted for his role.

In 1935 Russell urged Slichter to write the history of research at the University of Wisconsin, with stress on the role of WARF. "In view of the interest which you have taken all along, in the development of research, especially supported by our Foundation and the fact that you are originally the sponsor of this whole movement, it seemed to me that a very valuable record of accomplishment should be made more or less permanent." If Slichter had followed Russell's wishes we would now have a first-hand, nearly contemporary, lucid, but probably over-generous account of these matters.

Spending. One of the open questions was who should decide how money made available by WARF should be spent. It was probably fortunate that few foresaw the magnitude of the amounts that would be forthcoming, else there might have been a squabble for its control. Slichter and Russell were among the few who did. Russell would have been glad to see a large share of the funds assigned to the College of Agriculture. Slichter believed that scientists in other portions of the University should share in the benefits. Agriculture already had greater funds for research than the other colleges. Moreover, both because of the source of WARF money and because of the excellence of the scientists in the College of Agriculture, it was inevitable and right that agriculture would get a goodly portion.

The Articles of Incorporation, as described above, had stated that the purpose of the Foundation was to accept patents and other gifts and make use of the proceeds "to stimulate and promote and provide funds for further scientific investigation and research." The Foundation had power to enter into trust agreements, both as to the administration of patents and gifts and as to the use of the money derived therefrom. In assigning his patent rights to the Foundation, Steenbock had indicated his desire that the proceeds from his patent be used for research in the natural sciences; but having full confidence in the trustees, he had not legally specified this use in the assignment of his patent.

During Slichter's deanship almost all of the money received from WARF was used for the natural sciences, among which it was to be expected that an applied mathematician like Slichter would include mathematics.

The fact that Slichter conscientiously followed Steenbock's wish that the funds derived from his gift be used to support research in the natural sciences only—I believe that during Slichter's administration when exceptions were made they were with Steenbock's approval —does not mean that Slichter was unmindful of the needs of the social studies. He used the publication of Professor Frederic A. Ogg's report (1928) to the American Council of Learned Societies, "Research in the Humanistic and Social Studies," as an occasion to write a memorandum on the needs of research in the social sciences. He states:

~ The natural sciences seem to have outrun the social sciences—in other words, the power-creating sciences have outrun the power-controlling sciences. The problem is to stimulate progress in the social or power-controlling sciences. The job is to learn how to control our own power over ourselves—to direct the enormous social and economic forces let loose among us as we now control the far simpler fires, explosives, and electricities known to science. Many thinkers put this job first—higher in importance than all other jobs. The Ogg report lays down the facts and puts the problem definitely before us.

Research in the social sciences must be made a major effort. More men of outstanding ability must be called as scholars to the field, and they must be supported with ample funds for their work. The job is expensive, for the laboratory of the social sciences cannot be limited by the blueprints of an architect, for this laboratory is the world itself— industry, finance, government, are names of only some of the special rooms in this big laboratory.

He had spoken in the same vein in 1921 in his Phi Beta Kappa address, "The New Philosophy."

"Number 9" was used to balance the support given to the humanities and the social studies. Moreover, in the emergency actions, yet to be described, which were taken during the depression, young Ph.D.'s, some of them from fields other than the natural sciences, also were given annual appointments from funds furnished by WARF while they sought other jobs. Thus in 1932–33 some twenty-seven of these appointments were made, eighteen of them in the natural sciences, three in industrial processes, and six in the humanities and social sciences.

Moreover, when WARF supplied emergency funds in 1933–34 to place twenty-nine natural scientists from the faculty on research appointments for one semester each, it was possible to give seventeen similar appointments from "Number 9," five in the humanities and twelve in the social studies. If it had not been for WARF, these latter appointments would have been shared with the scientists.

As early as 1931, Slichter was bringing to the attention of the president of the Foundation the desirability of providing funds to help the publication of research in the social sciences. Beginning about 1960, some funds from WARF were also allocated to the social sciences, and, to a more limited extent, to the humanities. When the Foundation was initiated, it was far from clear how the money would be allocated within the broad range of the natural sciences. The legal authority rested with the trustees, but the qualifications that made them excellent trustees—including business experience and freedom from the details of faculty life, as well as the fact that they were men already absorbed in other duties—disqualified them from wrestling with and making judgments upon the intricacies of a research budget made up of many small and a few large items. Moreover, it is easy through a budget to interfere with the administration of the University and this the trustees were determined not to do.

The first allocation from WARF for research was a "grant-in-aid" in 1928–29 of $1,200 to Professors E. B. Fred and W. H. Peterson for their studies of bacteria. This grant was made in connection with a patent covering production of lactic and acetic acids, the funds being advanced to the Foundation by Trustees Hanks and Brittingham. Perhaps this marked the official start of Fred's long connection with the Foundation, though he had often been consulted by Slichter whose neighbor he was.

The precedent for future procedure was set in a letter of October 28, 1929, from Slichter as Chairman of the University Research Committee to Louis [Lucien] M. Hanks, at that time the Secretary-Treasurer of WARF, in connection with the 1929–30 grant. The letter is quoted in full.

~   Referring to your recent letter in which you have announced your support of the Research Committee in furthering research in the University for the current year to the amount of $5000, I will state that very careful consideration has been given to certain matters by the University Research Committee, and we have come to the following conclusions:

1—That $750 of the amount be spent to pay the salary of a half-time computer under Professor Ingraham of the Department of Mathematics to carry out the necessary computation and set up the mathematical formula for certain important research projects, especially under Professors Brink, Sevringhaus, Truog, and Tottingham. There has been nominated for this position Miss Beatrice Berberich who began her services October 1st and will continue to serve until August 1st, 1930 at a salary of $750. A committee consisting of Professors Brink, Jerome, Daniels, and C. J. Anderson, with Professor Mark Ingraham as chairman, has been appointed to consider applications for use of the computer so as to keep the work in control.

2—The Research Committee recommends that Professor Roebuck of the Department of Physics be placed on full time research for the second semester so that he may be enabled to make satisfactory progress on his important research known as the Joule-Thomson Effect. It is the understanding that the salary saved to the University by Mr. Roebuck's appointment will be used, first, to pay Mr. Osterberg, Professor Roebuck's assistant, $1000 for full time for the second semester, and, second, the remainder to be used for providing for the instructional work ordinarily carried by Mr. Roebuck in the second semester. This matter has been discussed with Professor Mendenhall and with Dean Sellery, and we understand this arrangement is satisfactory. Mr. Roebuck is to receive from the Research Foundation for the second semester the sum of $3000 as salary.

In accordance with these plans, I suggest that the sum of $750 be immediately put at the disposal of the University Board of Regents to pay the salary of the University computer and that next February the sum of $3000 be similarly placed to the credit of the Board of Regents.

A copy of this letter is being sent to the President and to the Business Manager of the University.

Three points should be noted: (1) WARF decided how much money was available. (2) It was to be spent on the recommendation of the Research Committee. (The Research Committee, under Slichter's chairmanship, was dealing with just such problems in connection with the expenditures from "Number 9," and it was soon routinely making recommendations concerning WARF funds. I believe that, under Slichter, this was an assumption of authority rather than a policy decision on the part of the University. In any case, it was a fortunate determination of function.) (3) Slichter made to the committee explicit recommendations which he was confident would be followed.

His enthusiasm for projects was infectious, but his forecasts were not always correct. He said the computer service "is of considerable importance, but involves very litle expense." The computing service has been of major importance, but the budget for the staff of that service, with its present ramifications, in 1968–69 was over half a million dollars, an increase of about 12% compounded annually.

As the above letter stated, the computing service was already in operation. But it had taken a little doing, for then, as at present, space was scarce. The only room in North Hall not in regular use was 401, in which Slichter's underground water materials were stored, as well as one of the early Michelson-Morley harmonic analyzers. As far as anyone had observed this sanctum had not been used for years, but friends such as E. B. Van Vleck, Herman March, and Warren Weaver, considerate of my well-being, advised me not to ask for this room— Weaver, in particular, saying, "If you value your skin [the word may have been 'hide'] you won't raise that question." But search produced nothing else, and with considerable uneasiness I went to ask the Dean for the space. He listened awhile, then slammed the top of his desk declaring, "That's what I've been saving that room for, for twenty years." (Mathematics had been in North Hall for only twelve years, but at times Slichter could be more precise than accurate.) The first computing service in an American university was installed within the hour.

As the funds of WARF increased, the primary investigation of all projects could no longer be made by the Dean, and some of this work was shared with various members of the committee, yet to a remarkable degree each item had Slichter's scrutiny and many his decisive backing.

Among the early and wise, but at times bitterly criticized, decisions of the Foundation was that in general the income from the patents would be used to create an endowment, and only the income from the investments of this endowment be used to support research at the University. If this had not been the case, there would have been a quick binge and a long headache. The University would perhaps have spent money too lavishly for a few years, and the contraction, when the first lucrative patent had run its course, would have been catastrophic.

But an emergency was building up. In 1925 the Regents, as already described, decided to no longer accept grants from outside organizations, such as the Rockefeller Foundation, and although this decision was rescinded in 1930, the flow of outside money could not be reestablished before the full force of the depression hit. The team of Steenbock, Russell, and Slichter again came into play. I believe that all three had supported the policy of spending primarily the investment income, but now they pointed out, as Russell eloquently stated, that, if the Foundation did not take drastic action to bolster science within the University, when the economy returned to normal there would be no worthy science to support. This was hard medicine for the Trustees to swallow. It is greatly to their credit that they did so, since it is particularly irksome to abandon a principle one has supported in the face of strong and unjustified opposition.

For the year 1932–33 the Foundation made a grant of $10,000 to give one-year research appointments to selected persons who had just received their doctorates at the University and had not been able to secure appointments for the next year. This was at the time when many institutions were dismissing young faculty members. However, at Wisconsin upon recommendation of the faculty, the younger staff members were retained and the salary cuts made necessary by doing so, like income taxes, took a proportionately greater bite from the larger salaries than from the smaller.

Independent of the depression, Slichter was in favor of some sort of postdoctorate support at the home institution for the better graduate students. On January 14, 1932, he wrote his son Sumner:

~ I am, however, inclined to believe that some sort of "interne" or "docent" would be a desirable innovation for our best students. When they get the doctorate they often are just in the midst of real work, with elaborate apparatus setup in science or elaborate bibliography in hand in other fields. Another year of work at the same institution might be most valuable. The National Research Council makes a mistake by encouraging its fellows to move to another institution. In most cases in science this is a mistake for the reasons just indicated. Next year we hope to hold as many of our best doctorates on the staff of instructors as we can, for we do not wish to have our best Ph.D.'s to be idle or take positions where they will be likely to go to seed. The depression may greatly injure some of the best doctorates, by forcing them to get work where they will be forgotten or buried alive.

Six months later he reported to Russell:

~ I am very glad to report that this plan has the hearty unanimous approval of the University Research Committee and myself.

Kindly note that the plan is very much more than a mere aid to the unemployed doctors. The plan is to make it possible for the best of our doctors to pursue further research for the following year. The students themselves as well as the University have already invested hundreds of dollars in the training of these young people. This training is perishable, and it will be a great social loss for their attainments to be per-

mitted to fade away during a long waiting period. On the other hand, a very small sum of money will greatly add to the social value of these people, and the small sum that it would be necessary to invest in their cases will have a large influence away beyond the value of similar funds invested at an earlier stage of their career.

The next year the grant was renewed, but the market for Ph.D.'s began to improve and no further emergency grants for this purpose were made. In 1933–34 a special grant of about $70,000 was made in order to pay the salaries on research appointments (mostly for a semester) of faculty scientists. A similar grant was made in the next year. In all, some sixty-one faculty members received such appointments. This move not only helped to retain these men and sustain their morale, but by relieving the University budget made it possible to aid other fields.

When one considers how many students and even faculty members strove to find tactful ways to avoid or mollify the wrath of "Hell Roaring Charley," it is amusing to see what a master of the art he was when dealing with possible donors, regents, or foundation trustees. He needed to be when he reported for the Research Committee that it would be desirable to continue to dip into the capital of the Foundation to pay the salaries of a number of professors in 1934–35. Judge Evan A. Evans in particular resisted the idea of doing something he believed was required only because of a dereliction of the Legislature in not giving sufficient funds to the University. He believed the State would continue to shortchange the institution if others supplied its needs. In reply to Evans's protest that he had not received the Research Committee's report in time to study it, Slichter wrote in 1934:

~ *Concerning the report and request I made to the Foundation, I would say that I sent a copy of that communication to George Haight as President of the Foundation on April 16, and at the same time sent a copy to Dean Russell. I did not think it proper etiquette to send copies to all members of the trustees, as it seemed to me that it would come more naturally to the trustees from the president.*

In reply, after asking about a few specific projects, Evans again protested this proposal. Slichter wrote him on May 17 giving details on various projects and adding:

~ *There is no doubt that the State Legislature should have been more generous to the university than it has been, although I appreciate that the difficulties of state finances can not altogether be ignored. On the other hand, the state has maintained here for sixty years an institu-*

tion that is among the leading ten or twelve in the United States by generously maintaining a large staff of able scientific workers. There has been developed here, as a direct result of this continuous and generous support, discoveries which are very profitable and useful, one of which has been turned over to the Foundation in trust. Perhaps it is not altogether unfair to expect that after continuous generous support has been given for a period of half a century to scientific investigations, that one of the discoveries that turned out to be unusually profitable should in an acute emergency be used for sustaining research at a time when the state for the first time in its history has found itself unable to perform its full duty.

This caused a minor explosion in which Slichter, Frank, and other faculty members were accused of thinking that the Foundation should merely serve the "needs of the hour." But Evans did agree to vote to approve the report of the committee. This agreement was the only sentence that Slichter marked in the margin. But on May 29 he again wrote the Judge:

~ I greatly appreciate your letter of May 26. I am quite certain that you will find that the results of putting a limited number of men of the staff on full time research for a semester has done a great deal to stimulate the spirit of scholarship at Wisconsin and to keep the morale of the men fully up to normal.

On May 1, I asked all members of the staff on leave this semester to send a brief report of their research problem. We have at present received replies from all but three persons. I am enclosing the second copy of each one of these reports which I hope you can take the time to read. I think you can see how grateful the workers have been for the unusual grant given to them so unexpectedly for the current year.

When you are through with these reports, I suggest that you turn them over to the Foundation's secretary, Mr. Russell, for his permanent files.

Slichter did not look up to persons of wealth or in high position, nor down on youth, but he did recognize his place in the power hierarchy, used his authority, and respected that of others.

The whole episode came after the basic decision had been in effect for one year, but the fact that emergency grants were dropped after another year may be due in part to protests such as those of Judge Evans. However, the practice of research assignments, so-called leaves, was now well established and continues to this day.

The support of new Ph.D.'s and young faculty members helped the

University through a major crisis, apparently did not decrease the support of the State, and put the University in a position where it could secure funds from other sources as these multiplied during the period of economic recovery.

The cooperation of Steenbock, Russell, and Slichter was essential at this time. Steenbock, through his generosity and foresight, made the Foundation possible; Russell was its executive officer; and Slichter was not only the chief force in guiding the Research Committee in the formulation of its recommendations, but also the channel through which the recommendations reached both the administrations of the University and of the Foundation, appearing before the trustees of WARF to urge these emergency actions.

Thus, as the funds in "Number 9" decreased due to the impact of the depression and outside foundation funds were cut off by action of the Regents, the amounts available from WARF increased. When Slichter became dean in 1920, "Number 9" provided $23,000 a year and in 1931–32 it had reached $75,000. From here it decreased during the depression to the level of $36,000 in 1933–34. By this time WARF was providing over $70,000 a year. We must remember that this was by no means the only money devoted to research. It was recognized by many that teaching must be based on sound scholarship and that in most fields the impetus from original investigation was a major stimulus to both scholarship and good teaching. The University's financial reports for many years included most faculty salaries under the heading "Instruction and Departmental Research." Moreover, there were research funds for specified organized research in several colleges, and these, in large part from federal sources, were substantial for the College of Agriculture.

However, as was mentioned earlier, there is something particularly invigorating about fluid funds allocated by a faculty committee to further the intellectual interests of scholars. Thus "Number 9" and the funds from WARF, dollar for dollar, had more significance than other research appropriations—even more than funds secured from outside, when both the securing of the funds and the accounting for them exhaust a significant portion of the energy that should go into research itself, for, in contrast, many felt buoyed by a conversation with Slichter about their aspirations and by the explanation of their projects to faculty colleagues on the Research Committee. The price in the time of committee members and administrators is real but the result magnificent. Moreover, during the period of Slichter's deanship, the amounts of money involved and the numbers of persons were not so large as to bog down the process. Now, at times, it would be a relief to have some fluid funds available in the colleges where some one person could say "Yes" or "No" as to their expenditures.

Slichter was a great dean. There were others, but not many, who would have been as wise a dean; still fewer would have been as vigorous. It would have been impossible to find another who to such an extent could have brought to life that which he touched. As an administrator Slichter should be the last man to copy, but perhaps the first to emulate.

Chapter 7   *Business and Civic Affairs*

Business

SLICHTER was not averse to cash, although in an undergraduate essay he had written, "I look at money from a distance and must be content to imagine and theorize about it!!" His barber told me that Slichter once had said that he had made up his mind to have $50,000 by the age of fifty—and had. This could not have resulted from placing in a savings account the residue unspent by a family of six living on a University salary. By the time he was fifty, his rapidly rising salary had reached $3,750.

His remunerative activities, in addition to teaching, for the most part consisted of (1) services as a consultant on questions connected with the flow of water, chiefly underground, (2) writing of textbooks, (3) erection and management of a business building on Carroll Street, and (4) investment in common stocks and real estate.

The first two of these were of a professional nature and were excellent contributions to geophysical engineering and to training in applied mathematics. These are described elsewhere; their business aspects are mentioned here. The third occupied a good deal of his attention, gave him an opportunity to engage otherwise unused talents, and was a picturesque outlet for a certain amount of native cussedness. It also augmented his income. The fourth allowed him to appear more of a gambler than he really was.

At no stage could one charge Slichter with shortchanging the University in order to make money. His research in connection with consulting was carried on at a high scholarly level and not only helped his teaching but broadened his interests and thus enhanced his usefulness when he became dean. His books, used here and elsewhere, gave a lift to the teaching of mathematics to engineers. His consulting was so extensive that for a man of lesser energy it might have encroached upon his university services, but it was done during vacations and while on formal leave. His wife might have had cause to complain (and did), but not his dean.

Consulting. Slichter's compensation as a consulting engineer of the United States Geological Survey in 1901 was $5.00 per day, and in 1903, $7.00 per day. This was quite a bargain, for his salary of $3,000 per academic year, computed on the basis of approximately 186 working days, would be equivalent to about $16 per day; and we know that his private consulting fee in 1904 was $20 a day. The charges to private companies had risen to $25 a day in 1909, and by 1911, to $50 per day—a figure he claimed had been his regular rate for several years previous to that date. By that time his salary was about $19 per day. He also stated in 1911 that his charge was $1 a day for consulting on a board of engineers, and $2 a day as an expert witness but that he would serve in that capacity only under subpoena. It was his policy not to charge for work done for Wisconsin municipalities or for public agencies in Wisconsin—work which, in a number of cases, was extensive.

Textbooks. Most of the senior members of the Department, up to the time that Slichter became graduate dean, wrote textbooks. Van Velzer did so and also collaborated with Slichter in the process. Professors Skinner, Dowling, and Dresden wrote college texts. Hart found writing and editing high school texts a lucrative enterprise. Slichter's first texts, written with Van Velzer, were several in the field of algebra and, in addition, a handy four-place logarithm table.

Although many men believe that they can write a better text than already exists, at least for their own students, this is seldom the case. However, Skinner's book on *The Theory of Investment*, Dowling's on the *Mathematics of Life Insurance*, and Slichter's on analysis were real contributions.

Slichter was concerned both with the quality of the books he wrote or edited and with the business details of publishing. No one seems to have questioned the teacher's right not only to a captive audience but also to captive customers. In 1912 Slichter wrote a prospective publisher arguing that a 10% royalty was too small. He said:

~ *It seems to me that the freshman and sophomore books should bear fifteen percent. The expense on my part of preparing these books is enormous, and yet we ourselves furnish a reliable and ready market for enough copies to eliminate all hazard from the publisher. In addition, the fact that the Freshman book replaces three books that the students have heretofore been obliged to purchase, makes it possible to get more than current prices for the book.*

Although Slichter was still getting some royalties from *Elementary Mathematical Analysis* at the time of his death, I would judge that the income that he derived from writing and editing texts, though not negligible, was not large. Colleges did not furnish a mass market as did high schools; nor were the sales of college texts as frequently controlled by state departments of education as were the sales of books in secondary and elementary schools, where a single decision could bring in orders for many thousands of copies.

Carroll Street Property. In 1900 Mrs. Slichter acquired a parcel of land 22'9" wide facing the Capitol on Carroll Street, between Mifflin Street and West Washington Avenue. In 1905 she bought an adjacent parcel 22' wide, southeast of the first. On these lots two buildings were erected which were rented as offices and stores; Spoo and Sons clothing store is now located in them.

In 1905 the Central Realty Company was formed, and Mrs. Slichter conveyed this property to it for "one dollar and other valuable considerations." This corporation soon became the Charles S. Slichter Company. Slichter acquired all of its stock. Its assets, including the above-described property, as well as land on the west side of Madison, were transferred to him in 1923. When he retired, somewhat over a third of his income was from rent. Throughout the whole procedure Slichter was the guiding force, although he did consult with others who had an interest in the project—in particular, Mrs. Slichter's brother John.

Of course the recording of the pleasing aspects of the construction, the prompt payment of rent, and the mutual satisfaction of lessor with lessee, did not call forth his expository talents, nor would it make interesting reading. In general the venture was successful but there were rough spots, and Slichter could make the most of them. A few sample cases follow.

The doors that were being furnished by the contractor did not satisfy him. On November 2, 1905, he wrote a letter of complaint to the architect who evidently was supervising the construction. In this letter he stated:

~ *These doors were rejected over two weeks ago, and the contractor is still talking about fixing them up. This has already caused a* loss of over two weeks time. On account of this dilly dallying and delay I desire to have the contractor informed that I propose to furnish the doors myself in conformity with the specifications and in accordance with the provisions of the contract. . . .

In order that there may be no loss of time or words, or misunderstanding to any parties, let me state that I do not propose to accept for the second floor of the new building the present doors, or any doors made out of them, or any doors furnished that do not correspond in the minutest details to the doors on the second floor of the old building.

These were not the only doors that gave him trouble. A vault door did not arrive on time. His letter to the dealer includes the following:

~ I am very glad that in this case you were dealing with parties that are capable of getting even with you, as we are frequently purchasing vault doors, and do not expect to have an experience of this kind again.

Slichter took pride in the "Slichter block" and naturally cared for its maintenance. He was rightly enraged when people failed to be careful in keeping the rest rooms clean. Part of his admonition said:

~ I am making every effort to keep the toilet rooms in the Carroll Block clean enough for fastidious people. It is impossible to do this unless all those that use this toilet room will try to cooperate. I suggest therefore that great pains be taken to throw matches and cigar stubs in the closets and not upon the floor, and that anyone having occasion to spit will use the closets and not litter the floor in the rooms. I have found recently that the space back of the closets has been used for this purpose and it has become quite disgusting.

Not as disgusting, but more amusing, were the occurrences that called forth the following protest—in this case, however, he must have taken some joy in preaching to an organization with which he had little empathy.

July 8, 1925

~ The Anti-Saloon League
20 North Carroll
Madison
My dear Sirs:
Complaint has been made to me that letters, waste material, waste paper, lighted cigarettes and so forth are thrown out of the third story windows and fall on the tin deck below and sometimes blow into the offices occupied by other tenants. Some of this stationery bears the

name of the Anti-Saloon League. Therefore I feel justified in calling
your attention to this matter in the belief that the difficulty lies in your
office.

Kindly note that there is a city ordinance prohibiting the scattering of waste material and that the lease of the Anti-Saloon League
specifically mentions the duty of tenants to conform to all city ordinances. As I understand it is the duty of your organization to take an
interest in law enforcement, I trust this important matter of the
observation of a city ordinance will receive your attention.

Anyone who now goes through Slichter's files is amazed at the business stationery of the period. Pictures of ugly factories, of which the
companies seemed to have been proud, abound. Robert Douglass,
owner of a prominent china and glassware store, listed twenty-three
important customers. William Owens, plumber and gas fitter, had a
picture of a then elegant lavatory of 1907. But all of these are surpassed
by the stationery of a realtor, L. E. Stevens, who occasionally found
potential occupants for space in the Carroll Street building. The top
quarter of the sheet shows Madison in 1905 as viewed from Lake
Monona, with two sailboats in the foreground and a small rowboat
near the shore.

On the back was a colored map of Dane County, with many modern names such as Shorewood and Nakoma necessarily absent, but
including such erstwhile places as Summit, about where the Verona
road (U.S. 18) now passes under the beltline, and Hyers Corners. On
the front, along the left, in very fine print, was the following description of the area, which gave a somewhat rosy-tinted picture of a region
where Slichter not only lived but into whose business activities he
entered. (It is true that you can count nine railroad lines on the map,
but they belonged to only three companies.)

~ DANE COUNTY, the second largest in the state in population, lies on
the southern and central part of the state and contains 1,200 square
miles or 807,400 acres. The county has the finest court house in the
state, a large hospital for insane, and other first class county buildings,
and owes no debt. The census of 1900 shows that the county contains
22,422 horses, 80,347 cattle, 62,000 sheep and 103,070 swine. The latest
crop reports show the production for the year 1903 of 2,359,651 bush.
of corn, 337,531 bush. of wheat, 3,669,517 bush. of oats, 885,275 bush.
of barley and 76,020 bush. of rye, 11,676,332 pounds of tobacco,
2,079,455 pounds of butter, 1,611,655 pounds of cheese, 181,000 tons
of hay, 311,181 bushels of potatoes, 306,190 pounds of wool, 51,507
pounds honey, 23,347 bushels of apples, 20,000 bush. of beans.

CITY OF MADISON. *Right in the center of Dane county, surrounded by four beautiful lakes, is located the city of Madison, the county seat of the county and capital of the state. Nine great railway lines lead out from the city, and within the city an extensive and complete electric railway. We will have two interurban railroads before snow flies, one from Madison to Janesville and one from Madison to Columbus and Fond du Lac. The University of Wisconsin owns 400 acres within the city on which land are college buildings costing $4,000,000. The faculty of instruction consists of 185 professors and instructors, and this year (1905) there are 2,870 students including 187 in the agricultural department. Besides the University library and city circulating library, the State Historical Society library (costing over $700,000) and the State library (law) are the largest in the west.*

The city owns a splendid system of water works and all water is supplied from artesian wells of a thousand feet in depth. The city and suburbs have a population of 25,000 and the county a population of 75,000.

The citizens have spent over $50,000 for pleasure drives.

For any information relating to Real Estate, etc., Address,

L. E. STEVENS, Madison, Wis.

Investing. Slichter seems to have enjoyed making investments. He superimposed speculation upon a backlog of conservative holdings in the stocks of well-established companies. He much preferred stocks to bonds. He also dabbled in real estate.

In 1908, a Mr. O. E. Hendel of Kansas City, Missouri, inquired as to whether Slichter would like to sell certain lots in Wyandotte City (state not given but probably in Kansas) and received a reply, "I regret to say that this property is not for sale, as I expect to start a glue factory on the same as soon as I can get funds." I cannot tell if this was a hoax, a threat, or a serious proposal—I guess, a mixture of the first two.

In 1911 he owned lots at the junction of Spooner Street and Rowley Avenue in Madison which he rented for the growing of potatoes, receiving one-third of the potatoes in return. He was still holding these lots in 1926, but certainly no longer as a potato patch one block from Randall School.

His vehement speech in business, as in the dean's office, sometimes offended people. Some were indignant, others hurt: For instance, "I am quite surprised that a Prof in the great University of Wisconsin, should exhibit his temper, and make the threats you did, over such a small matter." And in contrast: "I have been very proud of the opinion that you have had regarding me, and it hurt my feelings to have you think that I did anything unfair to you."

His longtime feuding with the Wisconsin Telephone Company is now perhaps as amusing as any of his business hassles. His first trouble with the company of which we have a record is described in a letter he wrote on April 27, 1909:

~ Since you have had exclusive charge of the telephones in Madison you have caused me great damage by your men getting on the roof of my building with spurs when making changes in the telephones. I called your foreman to the building and pointed out to him the damage done very recently. He explained to me the care that he usually exercised in the control of his men and so forth. I greatly regret to report that since the above conversation with your foreman, your men, in putting wires along the court, for one of the telephones in the building, again punched a number of holes in the tin roof of the court, which has caused damage to the Wisconsin Music Co. I enclose their letter. The damage in the hall due to your careless work during the winter has not yet been repaired. As soon as it is and I have the bill for the same, I will present the bills for these charges to you.

If you desire to restore the tinning in the corridor yourself rather than to pay for it, as above stated, you must do it immediately.

I suggest that your foreman give us his personal attention to the removal of the wires over my roof. It would be very unfortunate if we have a repetition of the experience of the past few weeks in punching my roof full of holes while removing the wires. You must be aware that leaks so caused are apt to do considerable damage.

This trouble was mild compared with that concerned with the service to his cottage on a privately owned line connected with the Madison Exchange of the Wisconsin Telephone Company. In the spring of 1911 he wrote to the Railway Rate Commission, which seemed to have also supervised telephone service.

~ My complaint is that this line is in such a dilapidated condition that even tolerable rural service is impossible over it. Many of the insulators on this line are no longer in place. In some places the connections to the supports have entirely given away, and in other places the wires show so much sag that every strong wind twists the wire together and cuts off the service. The joints connecting with the ground on this service are not soldered and every conceivable trouble has been experienced by me in the last year on account of these defects, and my complaint is shared by all users of the line.

He asked for an order to have the line put in workable condition and a reduction in charges because "in my estimation about three-quarters of the calls originating in Madison have not reached me because of the defective maintenance above referred to." The owners of the line made some changes but, according to Slichter, no real improvements. Moreover, he voiced a reasonable opinion that ten parties were too many for one line—eight should be the limit.

Some years later a party line to his city house was equally aggravating. The last letter we have about his telephone service was to the Wisconsin Telephone Company, dated March 12, 1934. It reads as follows:

~ *I desire to call your attention to the almost complete lack of service that we have at 636 North Frances Street, with telephone B. 3830. This telephone is on the same circuit with the number of a telephone which is in use for a half hour continuously at a time, so that our present service is valueless. We not only find that we cannot use our own 'phone, but we have many complaints from friends that try for long periods to get my home but find the 'phone busy at all times. Last Saturday I spent twenty-seven minutes waiting to call a taxi, because of continuous visiting on the line.*

I am sending you this notice in writing so that I may receive a definite reply from you concerning the action you will take. As the service is of no value at present, I take this way of telling you that I will decline to pay for any service unless suitable changes are made immediately. Otherwise, remove the telephone from the house.

If at times Slichter seemed a bit hard-boiled in business dealings, this was not his habitual attitude. There is evidence that during the depression he was considerate of his renters and worked out schemes, in at least one instance by profit sharing, to soften its effect upon them. Moreover, he had a strong sense of right and wrong. Thus, when the actions on the part of Robert W. Stewart, chairman of the board of Standard Oil of Indiana, were questioned, he not only sided with John D. Rockefeller, Jr., in condemning Mr. Stewart's actions on ethical grounds but also doubted Stewart's business effectiveness.

He must have been pleased by not only the scrupulous but the generous settlement made by his son Allen, one of whose earlier enterprises, in which his father and his brother Donald had invested, had gone sour. His father had put $5,000 in the project, Donald nearly $2,000, and Allen $100, but Allen wound up the accounts in such a manner that the loss to himself and to his father was $953.85 each, and Donald's loss was two cents greater.

For a man to whom business was of tertiary concern, he was success-ful. One might guess from this success and from his insistence on the prompt fulfilment of contracts and the prompt payment of rent that he was an exemplary bookkeeper, but one who has seen his files would have doubts. In fact, in business as in much else, it was his character and his intuition rather than meticulous care that brought him through.

Witness the following: On March 24, 1922, he wrote to his wife's niece, Bessie Byrne, "I have been trying to straighten out my records with you but have not found my memoranda concerning the notes." And further along in the same letter, this confession of confusion, "On Nov. 28, 1919, my check stub says '$1000 principal and 55 interest on $2000 note June 15 to Dec. 1, 1919.' Why this should say on '$2000 note' I do not know—do you remember any reason for that state-ment, and why interest should have been reckoned from June 15 to Dec. 1, 1919? Was this on the *old* note due your father?" It was evi-dent throughout that any doubts would be resolved in favor of Bessie. The niece was equally honest and also equally desultory—one of her letters remained unmailed for a year in the bottom of her shopping bag.

There were also financial mix-ups with his side of the family. In 1910 his brother David wanted to know if Slichter had paid in full a note of $200.00 of January 26, 1891, bearing 12% interest on which are recorded payments totalling $142.50. We do not know the out-come, but Slichter became his brother's executor.

Business was rather pleasant, for income is nice and fuming exciting. But few persons as immersed in the present as Slichter, desired as deeply as he to look at things *sub specie aeternitatis*. In 1911 he pre-sented to the Madison Literary Club a paper on "Industrialism," which gave his thoughts on whither business was leading America. It starts with the eternal refrain, "It is only rarely nowadays that anyone writes hopefully of our own times." Further along he related science to industrialism:

~ *In a large sense, science and industrialism are not two forces, but a single force. Industrialism is merely science in action, or militant science. But this distinction is a large one. To make industrialism from science, one must add other elements—such as ambition for power, greed for exploitation, or lust for money, or some combination of these. Of course industrialism could not have developed except from the soil of science.*

                                        .   .   .

If science has given us the tools, the methods, the point of view, industrialism has given us the laboratory and the fiery furnace in which to test them. The assembling of men in great dependent groups, the subdivision of human effort, the new conditions of life, the accidents and dangers of modern industrial employment, have forced upon us problems in bulk, and not in single instances. The business world has shown us how to divide up investments, risks, and profits by the joint-stock organization. It has drilled us in the elimination of hazards and the division among the many of the ownership and rewards of industry. This very phenomenon emphasizes by contrast and makes it inevitable that society share the hazards of the life of the individual. To place the burdens of the individual upon the broad shoulders of the state is therefore but a reflex from industrialism itself. A community of interests among the prosperous classes and class hatred between the proprietary and the working classes cannot permanently coexist. If the industrial trust brings peace where there was war, this peace must finally extend to humanity itself. Industrialism has eliminated the middle ground and the possibility of compromise. Peace between the giant groups is progress; warfare between the giant groups is destruction. Science cures the ills it itself creates.

When, in 1938, this essay was published in *Science in a Tavern*, he added a three-page note which includes:

~ Thus far to 1912, when the above paper was first published. This is the year 1938. The optimism it reveals will now cause a smile, and the further fact that I still adhere to it will raise a laugh. But the forces let loose in society by industrialism have not yet compounded their major resultants. Unfortunately evolution is in no hurry—the results that man impatiently requires in a few weeks, nature is apt to fumble over for years or even for ages. Nature seems to possess no sense of economy of time. She devoted a hundred million years trying to make a citizen of the trilobite. . . . No remnant, not even a remote descendant, of this race of creatures now inhabits the earth. Nature is prodigal of time—one hundred million years allotted to the trilobite experiment, and only a few million in all to the life of homo sapiens. No wonder, then, that man becomes impatient. . . .

. . . The examples from the industrial age, feeble as they may appear in the present world chaos, nevertheless seem to show that cooperation is slowly emerging as the only philosophy on which man can survive. Science cures the ills it itself creates.

A clear echo of this was heard on May 3, 1957. In that year the College of Engineering of the University of Wisconsin dedicated its annual Engineers' Day Dinner to the sons of Charles S. and Mary L.

Slichter, and the eldest, Sumner, gave the address entitled, "Technology and the Great American Experiment." Near the end he said:

~ An outstanding contribution of technology, and of science as well, to the values that guide our lives is their influence upon the sense of social responsibility. Strangely enough, this effect has escaped the attention of the moralists who worry about the effects of technology upon our ethical standards. Before the rise of modern science and technology, man lived in an environment which was pretty much beyond his control. When he had tough problems he could do little about them. But as science and technology have given him insights into his environment, this feeling of helplessness has been slowly disappearing. Today, we realize that even if we cannot solve a problem completely, we can at least do something about it. The result is a growing sense of social responsibility—a growing disposition to make planned attacks on our problems. No longer can we ignore human suffering or misery by taking refuge in the comfortable excuse that conditions are beyond our control, and that all we can do is to provide charity. Thus, by giving us new powers, science and technology also give us new responsibilities and foster a new willingness to act on those responsibilities. In this way science and technology contribute to the advancement of morals.

It is a fortunate man whose son repeats his clarion call.

Civic Activities

This section is placed in the same chapter with Slichter's activities as a businessman; otherwise, some very difficult allocations would have had to be made. Clearly, service on the city council is a political activity; but how about defending the shoreline near one's own property or urging the asphalting of the streets around the Capitol, but not before the proper sewers had been installed? We live in a continuum of which time and space furnish only the four most measurable dimensions.

Politics. No man as alert as Slichter could be oblivious to politics, yet it was not one of his major interests. His picturesque speech might easily have made him a noted partisan, but although individual issues called forth his energy, party did not. He found congenial, diverse groups of politicians. For instance, governors—he was on good terms with the La Follettes, with Kohler, and with Schmedeman. He got along well with Senators Vilas and Spooner. He dealt easily and often with political figures and perhaps enjoyed even more the company of jurists, especially Justices Burr W. Jones and John B. Winslow of the Supreme Court of Wisconsin.

To some extent he was an anti-chameleon, always in contrast with his background; but with such men it was contrast rather than discord. This statement is not without exception, for he strongly disagreed with Senator La Follette's war position and signed the famous "Round Robin" condemning the Senator.

The Common Council. Slichter's one experience in elective political office was as a member (from April 1897 to April 1899) of the Common Council of Madison. He ran as a Republican in the Fifth Ward and carried both precincts with a total vote of 417 to his Democratic opponent's 356. He had had the support of the *Wisconsin State Journal* which, after speaking of his youth, his promise, and his Madisonian wife, concluded: "It is quite the practice in German cities to choose University Professors to the Municipal Council and Professor Slichter should have the vote of all citizens of the Fifth who think none are too good for that body."

Whether the *Journal*'s support was a benefit or a handicap is not clear, for it supported the total Republican slate; however, Matthew J. Hoven, who was elected mayor, was a Democrat, as were a majority of the Council. It is interesting to note that Mayor Hoven (who was defeated for reelection in 1898 but regained the mayoralty in 1899) was the grandfather of John Pike, now Director of the State of Wisconsin Investment Board, and the husband of Ann, the daughter of Slichter's son Donald.

Slichter was appointed to the committees on finances, ordinances, schools, cemetery, and street lighting. Considering his affection for dogs, it is too bad he was not on the committee in charge of the pound —a committee which seems to have disappeared in 1898. The *Proceedings* of the Council from 1897 to 1899 are interesting reading, more because they give a picture of the city at that period than for the light they throw on Slichter. Slichter's attendance was regular; and most of the votes of the Council were unanimous and on the few that were not, he usually adhered to party lines. The salaries of city employees and the conditions of the streets, sidewalks, and crosswalks were among the chief items considered. Streets, such as Mills from University Avenue to Dayton Street, were approved for macadamizing during this period. Slichter succeeded, in August of 1898, in attaining authorization for brick crosswalks on the north side of Langdon Street, where it crossed Lake Street and Park Street. These were presumably necessary because of the mud and pollution from the then current mode of transportation. As to salaries, at the April 22 meeting in 1897 the salary of the Chief of Police was set at $15 per month and of the other policemen at $10 per month. This is not to be taken as a judgment as to what should have been a standard scale for policemen; the rates were set perhaps more to secure their resignations than for any other reason. The resignations were later requested directly, on a vote which adhered to party lines.

On October 16, 1897, Slichter tried to have postponed the decision in regard to a sewage plant until an analysis of a contracting company's profit could be made. His motion lost 12 to 3; and on the original motion, his was the only negative vote, though one other alderman refrained from voting. After an analysis of the various methods of sewage disposal, he made the following interesting and modern statement. It must have been a real annoyance to have expertness thrust aside so bruskly.

~ *What do we want a sewage-disposal plant for? It is not to protect our public health or our water supply. It is to protect our lakes—to attempt to restore them to their original condition—to keep down the present enormous growth of weeds.*

To show the original condition of the lakes about Madison before they were polluted, note the following in quotation from the description of Wakefield, who traveled through this country and described it, July 20, 1832:

"Here it may not be uninteresting to the reader to give a small outline of those lakes. From a description of the country a person would very naturally suppose that those lakes were as little pleasing to the eye of the traveler as the country is, but not so. I think they are the most beautiful bodies of water I ever saw. The first one we came to (Third lake) was about 10 miles in circumference, and the water as clear as crystal. The earth sloped back in a gradual rise. The bottom of the lake appeared to be entirely covered with white pebbles, with no appearance of its being the least swampy. The second one we came to (Fourth lake) appeared to be much larger. It must have been 20 miles in circumference. The ground rose very high all around, and the heaviest kind of timber grew close to the water's edge. If these lakes were anywhere else except in the country they are they would be considered among the wonders of the world. The other two lakes (First and Second lakes) we did not get close enough to for me to give a complete description of them, but those who saw them stated they were very much like the others."

The effect of the settlement of the country and the putting of sewage in the lakes, has been the enormous increase in the fertilizers in solution in the lakes. This has disturbed the balance originally existing between the animal and vegetable life in the lake. We should restore this original balance as far as possible by adopting land treatment and using up the fertilizers in cropping the soil.

A great mistake is made in supposing that dilution on a large scale takes place as soon as sewage reaches the lake. As a matter of fact, about all the dilution that takes place, takes place where the effluent reaches the Catfish. If the dilution is one to five or one to ten, then the dilution in the lake is one to five or one to ten, unless an enormous weed growth keeps it down. The lake is but a reservoir in which the Catfish expands, the inflow from springs and the drainage area of the lake probably not much exceeding evaporation.

In conclusion, permit me to say that as good an effluent as will come from the new scheme could be obtained from a simple coke strainer, which could be built for a few thousand dollars, and operated at a lower cost than present scheme. Sludge could be effectively disposed of by burning coke.

Slichter held this opinion for a long while, since the minutes of the meeting of Town and Gown on November 13, 1943, say: "A meeting with Adams May 9, 1896, recorded that Slichter and Van Hise opposed a proposed chemical system of sewage disposal for Madison. Slichter still opposes the system which eventually was tried and abandoned."

The lakes were his constant concern. He was active in protesting the dumping into the lake of sewage from the Wisconsin State Hospital at Mendota. He participated in passing an ordinance against polluting the lakes and forbidding filling them in.

During the second year of his service, one of the major topics for consideration was the possible establishment of the Madison General Hospital. The Madison General Hospital Association wanted a green light to go ahead with establishing a hospital and asked the city to subsidize it to the extent necessary but not to exceed one hundred dollars a month. The city offered half this amount. The Association then made a counterproposal that the city contribute a single sum of $10,000 toward the building of the hospital, the sum to be covered by an issue of bonds. Slichter pushed for the granting of this request, but it was turned down. Mayor Whelan, who had succeeded Mayor Hoven, opposed the proposition on the ground that public funds of that magnitude should not be used to endow an enterprise unless there were a greater degree of public control than was indicated. Medicare had a hard time aborning even in Madison.

Slichter was on the losing side of another issue, namely, the granting of liquor licenses in the Fifth Ward bordering the University. Slichter's opposition to the licenses carried the committee, largely on the basis of the request of the University authorities; but the Council did not support the committee and, in fact, went on a binge of license granting. Slichter voted with the minority in these cases.

As mentioned above, Whelan became mayor in 1898 and Slichter was put on a different set of committees, becoming the chairman of the Committee on Claims and also on the important Committee on Streets and Sewers. At the last meeting of the old Council in April of 1898, Slichter was requested to speak for the group in presenting to Mayor Hoven a gold-headed cane in appreciation of his services. His grace in doing so was particularly noted in the Council minutes.

Slichter did not run for office in 1899, thus ending his political career at age thirty-five.

The engineering problems of the city, in particular paving, smoke, and traffic, were of interest to Slichter.

Paving. In 1905 there was a minor turmoil concerning the condition of the streets around the State Capitol. The governor and legislature wanted an asphalt pavement. (Presumably the pavement was macadam, as was that of many other streets.) After hearing how irritated Governor Davidson was with the city, Slichter wrote, on June 13 to Colonel Vilas, describing the situation and saying, "I do not feel justified myself in doing anything that could be construed as delaying the carrying out of the wishes of the state legislature."

This, however, did not mean that the interests of the property owners should be overlooked. When Slichter received from the city clerk a notice of special assessment for asphalt on Carroll Street, he replied, on September 1, 1905:

~ I do not plan to put any obstacle in the way of this improvement if proper provision is made for a new sewer on the above street before the construction of the asphalt pavement is undertaken. I called the attention last spring of several of the aldermen and city officials to the necessity of constructing a new sewer on Carroll Street before the asphalt is put in place. The present sewer is entirely too small and is at so shallow a depth as to be useless to the owners of property on the street. If it is planned by the city authorities to cover this sewer with asphalt pavement, with the information at hand that I have given that the present sewer is entirely inadequate, I propose to put every obstacle, legal and otherwise, that I can in the way of the consummation of such a blunder. Therefore before signing the notice that I have received I would like to know whether or not any action has been taken by the city authorities in this direction. I shall be much obliged if you will give the Mayor and City Council formal notice of the condition of affairs as they exist on Carroll Street, and the imperative necessity for prompt work in the construction of a proper sewer, before the street is paved.

Smoke. As to smoke, his chief concern was that the heating of his Carroll Street property cause as little smoke as possible. He wrote a memorandum concerning the care of this building during his absence in Europe in 1909–10, with special instructions for tending the furnace. He was insistent, not entirely successfully, that his directions be followed. Someone, perhaps Mrs. Slichter's sister Isabelle Byrne, kept a record of the level of smoke at five-minute intervals for three days in February 1910—a more patient process than Slichter would ever have endured; but he preserved the record.

Traffic. In 1928 it was proposed to put University Avenue in Madison through to West Washington Avenue in order to provide an arterial route from the center of the city to its western residential districts.

The following letter to Mayor Schmedeman, written May 14, 1928, is an able, if somewhat didactic, exposition. The mayor had already been Minister to Norway and was to become Governor of Wisconsin, was a fellow Episcopalian with whom Slichter frequently worked, and had helped him in connection with the effort to arrange an extended visit by Einstein to Wisconsin. We do not know what effect this letter had; however, its negative advice was in accord with later lack of action, and its positive advice, considering the effect of diminishing rail traffic, not completely different from later developments.

~ I have been much interested in your statements concerning the opening of University Avenue to West Washington Avenue, and I trust you will be glad to receive from me the observations that I have made that seem very pertinent at the present time. I am sure that the development of West Washington Avenue as the main artery to the center of the city is beyond controversy. The matter that I emphatically doubt is the opening of University Avenue to that street. I am quite sure myself that when that street is opened and the city has gone to all that trouble, you will be greatly disappointed, and in the years to come it will be thought that your administration has not given careful consideration to the development of Madison traffic. The use of University Avenue as an artery violates two of the well established rules now known to govern city traffic. One of these rules is that a feeder directed away from the central zone should feed the residential district with a minimum of left turns. The other rule is that in directing traffic to the center of the city it is immaterial whether the arterials are fed by left or right turns as this matter is controlled by stop signs.

Applying these principles to University Avenue you will note that the region of University Heights and Wingra Park must be fed by left turns. Going west on State Street there are left turns at Johnson, at Gorham, at Gilman, and Park Streets and again at Mills Street, Warren Street [Randall Avenue], and Breese Terrace and Allen. This means that nearly all the residential district is fed by left turns, making the traffic very difficult and the opening up of University Avenue to West Washington Avenue would only emphasze the interlacing of traffic and the difficulties mentioned.

The obvious way of meeting this difficulty is to use West Washington Avenue to the Chicago, Milwaukee and St. Paul tracks and build a viaduct to Regent Street which would permit the feeding of practically all of the west end traffic by right hand turns and would make an excellent thoroughfare to South Madison. This would reserve University Avenue for the through traffic to Middleton and the fraction of the traffic required by the university and would take all the burden off of State Street that is needed. The construction I propose is expensive but when once done, the traffic difficulties of Madison will be taken care of as well as can be and we will have gone at the problem in a metropolitan fashion and not in the confused way appropriate to a small town. For an improvement so important and crucial, bond issue would be justified. All other schemes such as the building of a crossway across Monona Bay could be postponed for a considerable length of time, as the viaduct at West Madison railway station would enable the Bay road in Brittingham Park to be used to the limit. Kindly note that the Bay road will feed a large area of country with right turns only.

I trust that before serious mistakes are made that this problem will be taken up in a more serious way than it has in the past and that something can be done that will be a lasting memorial to your wise city administration.

Grace Church. I trust that the inclusion, under "Civic Activities," of one episode in Slichter's relationship with Grace Church will raise no constitutional questions.

Grace Church is located almost adjacent to the Slichter building on Carroll Street and for many years was the only Episcopal church in Madison. Slichter was a frequent but not a compulsive attendant, rented a pew, and made regular contributions. From April 1906 to April 1908, he was a member of the vestry. During this period, on April 3, 1907, Walter Noe, a warden of the church, wrote Slichter (and I presume other members of the vestry), suggesting that the church pay the rector's coal bill and stating that the treasurer's report indicated that this was financially feasible. Slichter's reply, written on April 5, follows:

~ I have your letter of April 3, concerning a matter that you intend to bring up at the next meeting of the Vestry of Grace Church, which you state is the proposition to assume a bill of $240 for coal used by the Rector at his residence.

This matter was mentioned to me some time ago, and I am personally strongly convinced in my own mind that it is a very great mistake to go into matters of that kind. I am quite willing to consider raising the salary of [the Rector] now, or at any time that the finances of the parish will justify the same. If our finances justify the paying of the coal bill, they will certainly provide for a raise in the Rector's salary.

My objections in the matter are both specific and general. First, I think it is exceedingly undignified for us to take up bills for the Rector and pay them; I feel sure that he would very much prefer to have the amount come as an increase in salary. The payment of the bill involves knowledge on our part of his private affairs that we have no right to go into. Moreover, I regard it as an exceedingly bad business principle to take up the payment of any of the Rector's bills. If we establish a precedent by paying his coal bill at this time, the result is apt to be that in the future the Rector will become careless in the exercise of economy in the matter of fuel, and will not try to purchase the same at the lowest price possible or use as cheap a fuel as the plant can handle. We will then have bills of unknown amount coming in to us which we cannot estimate in advance. Any addition to the Rector's salary we can make definite in the beginning and the funds so given him he can use for fuel or for any other purpose that he sees fit. I personally see no more reason for paying his coal bills than for paying his doctor bills, or electric light bills, gas bills, or his water rent. In all such cases very much more economy will be secured if we pay the Rector his salary and let him spend it as it seems best to him.

While I have the greatest admiration and sympathy for [the Rector], and am willing to do anything that the parish finances will justify in the way of increasing his salary, I can not go so far as to take up his private accounts and assume them as part of the burden of the parish. I trust that on further consideration you will come to the conclusion that a different way will be the wisest, both as a precedent and as a business principle, for the handling of these matters.

Would that more parishioners were as wise and understanding!

Chapter 8 ᵒᵛ *Social Life*

S LICHTER was addicted to clubs. He read about them, he wrote
about them, and he joined them. His role in founding the University Club is described in Chapter 5. There were two clubs, the
Madison Literary Club and Town and Gown, which were part and
parcel of his very being.

The Madison Literary Club

The Madison Literary Club was founded October 8, 1877. At most
times it has had about fifty to sixty members. Their spouses are co-
members (in a few cases members in their own right); those who no
longer wish to give papers become honorary members, while widows
and widowers of members remain in the Club. Slightly over one-half
the membership has usually been from the faculty of the University
of Wisconsin, largely but not exclusively from the humanities. The
remaining members chiefly are lawyers, editors, and ministers, with a
sprinkling of others. The Supreme Court of the State has been well
represented.

Except in summer, there is a meeting on the evening of the second
Monday of each month. After a half-hour for gathering and chatting,
the program starts at eight o'clock. There is a paper by a member (the
norm, forty minutes); then two discussants (ten and five minutes the
goals, often exceeded, especially by invited nonmembers). A general
discussion follows, ranging from minimal and dismal to lengthy and
lively, after which nonalcoholic beverages and some trivia in the way
of cookies or token sandwiches are served. Some meetings are dull.
(My first paper, when I thought subtleties of mathematics could be
understood by faculty members and their wives, was not really rescued

even by the discussions of Professor McGilvary and Dean Slichter.) Generally, however, these have been fine evenings. It has been an admirable group although there has been more than a slight tendency for the Club to take itself seriously. Slichter became a member during 1901–2 and remained an active member throughout his life, reading nine papers, as listed in Chapter 9.

Town and Gown

Clearly Town and Gown meant more to Slichter than any other club. A group of articulate men, each of whom was accomplishing much; a varying group, appreciative of diversity, fine food, fine drinks, and fine conversation, these members found joy in each other's company.

Town and Gown coagulated in 1878, the founder (according to Birge) being Burr W. Jones, a Madison attorney, later to become an Associate Justice of the Supreme Court of Wisconsin. The three who first decided that there should be such a club were: Burr Jones; Charles Noble Gregory, another Madison attorney; and David B. Frankenburger, professor of Speech at the University of Wisconsin. Among the original group invited to join with them in forming the club was E. A. Birge. The roll of the Club, as signed in its annals in October 1894, consisted of (in the order of signing): Burr W. Jones, Charles Noble Gregory, David B. Frankenburger, Edw. Asahel Birge, Edward Thomas Owen, Joseph Jastrow, Charles Reid Barnes, Frederick J. Turner, Lucius Fairchild, Charles Kendall Adams, Charles Richard Van Hise, and Breese Jacob Stevens. Charles Forster Smith first signed the scroll on March 5, 1898; Charles H. Haskins on January 12, 1901; and on April 5, 1902, three new members: John B. Winslow, Chas. S. Slichter, and Moses Stephen Slaughter. From that date until his death, Slichter attended the meetings whenever possible.

A word of introduction concerning these men is appropriate. All of these seventeen belonged also to the Madison Literary Club. The presidents of the University of Wisconsin, Birge, Adams, and Van Hise, appear throughout this book. Owen, who was a professor of French at the University of Wisconsin, has already been introduced in connection with the University Club. Jastrow was an eminent psychologist, also with the University; Barnes was a professor of Botany, but had left for the University of Chicago before Slichter joined the Club; Turner, the historian, was the Turner mentioned in "The Diary of a Student," quoted in the description of Slichter as a teacher; Fairchild was an ex-governor of Wisconsin; Stevens, a lawyer, and for a long period a Regent of the University of Wisconsin. Smith was a professor of Greek, Haskins, of History, and Slaughter, of Latin at the University. Slaughter was also a neighbor of Slichter's; while Winslow, at the time he joined, was an Associate Justice of the Supreme Court of Wisconsin and a few years later became its Chief Justice.

I shall not list here those who joined during Slichter's membership but some will be mentioned later. They included three University Presidents, two Governors of the State, two Justices of the State Supreme Court, and one United States Senator, as well as an assortment of professors.

The Club met on Saturday evenings, approximately every other week during the academic year, at the homes of members. On each evening the Club first imbibed, next consumed, and finally the members enlightened each other. Usually someone led the enlightenment but the conversation was always general and sometimes only general. Each host wrote a one-page account of the evening in the annals of the Club. The reading of these must have added joy to the meetings. The fact that the host was the recorder means that less immortalization was provided the menus than they merited.

The Club deserves a chronicler, but here I shall try only to give the flavor of the meetings revealed either through references made to Slichter or through excerpts from the minutes which he wrote. In general since he, more than any other member, poked fun at the others, they reciprocated with many thrusts at him.

The minutes of April 1, 1905, are quoted below:

~ The Club met with Jones, 112 Langdon St. April 1, 1905; present Frankenburger, Jastrow, Slaughter, Slichter, Turner, Vilas, Van Hise, Owen, Winslow, Jones. Birge in Chicago; Smith still in Europe. Slichter had just returned from Denver where he had just been giving testimony as an expert witness on the celebrated case in which the State of Kansas is seeking to prevent Colorado from diverting the waters of the Arkansas and other rivers for irrigation purposes. He commenced by saying that he could tell the story in twenty minutes. Turner and Slaughter at once looked earnestly at their watches. Slichter stated the contentions of the States and of the United States, which has become a party, and as far as we could gather it is claimed by Colorado and the U.S. (1) that there has been no diversion at all; (2) that it is very slight; (3) that the diversion is a great boon to Kansas. He talked of the effect of the rivers upon the underground waters of the adjacent regions and for a time the discussion was quite serious. Slichter evidently thought he was still under oath. Then he

tried to make the occasion still more impressive by telling his ad-
ventures and hardships on a trip into the mountains 12,000 feet high
on Moffat's new railroad. The Expert on Moisture easily passed from
underground waters to a blizzard on the dizzy heights where he was
blockaded. The climax was reached when he told us in the most
thrilling manner how he had survived twenty-seven hours with nothing
to eat or drink. The rest of the evening was anti-climax. Although we
felt that we ought to sigh, somehow we laughed. The fact is, we had
all seen Slichter eat. We had entertained him. We had paid the bills.
There was an irrepressible desire to know the details of Slichter's fast
and how it ended. For two hours he was the storm center of such a
cross-examination as not even the Expert or the Town and Gown had
ever experienced. His twenty minutes must have seemed to him like
twenty-seven hours. The denouement finally came and it developed
that the miraculous fast was relieved by a copious supply of pie, peach
pie, and beer. There were numerous apologies to Slichter by members
of the Club for their undue levity and seeming incredulity, and Owen
formally apologized for the whole Club. But Slichter returned no thanks
and the apology was taken back.

Slichter's hardships and perils in the mountain fastnesses called up
many other reminiscences and the Club adjourned proud that they lived
among heroes and in a heroic age.

Burr W. Jones

The diversity of topics which the Club discussed is too great to be
represented by a few instances. Moreover, the reader must be warned
not to take as fact what one member is said by another to have reported, unless he can at times believe the incredible. The waters of
the Arkansas River are still the subject of dispute.

Throughout this biography there are numerous quotations from
minutes of the Club about Slichter's opinions or actions. The following two remarks in the minutes, both by Birge, illustrate the conviviality and intellectual contributions which Slichter brought to these
meetings. The first is from the minutes of December 15, 1917; and the
second from those of January 11, 1919:

~ Slichter is clear in his judgment that the power of research is cor-
related directly with the capacity for food and drink, & that candidates
for the R.S. [Royal Society] should be rated according to their ability
as trenchermen. To this doctrine the Club subscribed nem con.

Adjourned about 11.

And, "Slichter talked both wisely & interestingly on institutions for research."

The fun that Slichter had at these meetings is shown in the following excerpts from minutes that he wrote. The first is from June 2, 1923:

~ Pirates, however, were not like the pirates of the dime novel or of the comic opera, but a miserable lot much like a cross between a college professor and a cowboy.

The following solemn report is from the minutes of September 26, 1931:

~ The club statistician furnishes the following facts: This is the 53 yr. and 901 meeting; and Warren Weaver, admitted this night, is the 33rd member. We have spread 700 pages of minutes and consumed 17⅔ barrels of intoxicating liquor or a mixed average of only 7/11 barrel per member. We have eaten 36,234 lbs. of solid food at a cost of $27,642.98 or $1531.39 per ton, not counting laundry, breakage or overhead. This food, if properly ground and stuffed in hog gut would make a sausage extending from Madison to Pokerville, or, if stood on end, would reach from sea level to one mile above the stratosphere.

The third comes from the minutes of October 20, 1928, the fiftieth anniversary meeting:

~ That there may be no future misunderstanding, it should be here recorded that our little club has always dwelt in the Pale of Sobriety—in fact it was founded in the very desert and central reaches of that land. But an evening on the marginal shores of that isle has its delights. The view seaward is soothing and enchanting, even though one or two of our Club in their caution turn their backs upon its liquid beauty. Of such as these is the Kingdom of Abstinence. . . .

But the writer finds that he cannot put in words the record of this meeting. The true session was a communion in things unsaid and made up of a spirit present in all of us, but passing only subconsciously from each to each—surely all the better understood because unworded and unwordable.

It is interesting that Senator Vilas discussed in 1906, as described in minutes written by Slichter, the need of research professorships at the University of Wisconsin. At that time he had already provided for them in his will, but it would seem had not let even his cronies in Town and Gown in on the secret.

~ Col. Vilas delighted all with his large views of University work. Ever since a young man he had believed that the University should have a few endowed professorships, attractive to the very best scholars, the holders of which could devote their lives primarily to productive work.

Slichter's deep affection for his fellow members is made clear by memorials which he wrote to two of those who had died:

IN MEMORIAM

~ On April 5, 1902, Moses Stephen Slaughter, together with the present writer and another dear friend, signed the book of the Town and Gown. It would be fitting indeed if I could here record a true appreciation of the years he spent in this intimate circle, which meant so much to him and so much to us. It was the joyousness of youth that he brought to us in his boundless interest in the world of letters, a joyousness and an interest that seemed as fresh and as ardent as those of a college lad who is surrendering to the great masterpieces for the first time. Perhaps this, after all, was the central fact in his charming character, and the explanation of his extraordinary success as a teacher and as an intimate friend. He never out-grew the enthusiasms of his college days. Someone has said that "college life makes up the last chapter in the Book of Youth." Most of us slowly thumb those golden pages and then reluctantly close the book. But Professor Slaughter was one of the gifted few—he did not close the Book of Youth. By a power that came from his affection for things of college life, he grasped still some of those glowing pages—he could not close the Book of Youth. And if amidst the cruel blows he was called upon to bear, his physical nature did in part give way, such blows could have no power over his great spirit, except to refine it, and to make it give forth more abundantly to his students and his friends.

His days of suffering were but few and on the last day he had peace. I cannot think of him as less than willing to view that vast starry night. Perhaps he could say:

> "O Death! I am thy friend,
> I struggle not with thee, I love thy state:
> Thou canst be sweet and gentle, be so now;
> And let me pass praying away into thee
> As twilight still does into starry night."

IN MEMORIAM: FREDERICK JACKSON TURNER
Nov. 14, 1861—March 14, 1932

Among the three or four tutors who in 1886 sat in the marginal seats at faculty meetings in old North Chapel, there was one at least whose romance in scholarship had already begun; one, who even then, was in that fervent and unrestful activity that marks the apprentice years of a scholar. Like Van Hise, Frederick Jackson Turner was wholly a Wisconsin product. He was born at Portage when it was still the center of many Indian villages and for fifty years had been a focus where the red and white converged. In his university he had won the art of public speaking and had been trained in the peculiar system of Wisconsin debate. Here President Bascom illumined for him the problems of human life. Here Professor Allen traced with him the philosophic threads of history. Here the lakes and streams, the prairies and the wooded North seemed to make natural causes merge with human and political and historical causes; indeed, seemed to frame history into a sequence of physical events. From Professor Allen came the revelation of the forces generated by the impact of the pioneers upon the margin of the wilderness. This to Turner became the passion of his life. He was to interpret the American frontier as a power in history—not as the periphery reacting upon the center, but as the spirit of America reacting upon and throughout the corpus.

In 1890 Turner obtained his doctorate at Hopkins. At that place many lifelong associations were formed and he returned to a professorship of American History at Wisconsin fully ready for its duties and convinced of the richness of our local collections. Here he attained the ideal longed for by every scholar: To found a school of thought and to train in it the first ardent disciples of that school. But in 1910 came the call to Harvard—requiring a decision hard to make and sure to be partly wrong however made. He had sorrows to escape—not to escape but to mellow and perhaps that must decide. He made the change—it is uncertain even now whether the influences might not have been greater at the western and more native center. But he found many new friends of note, and appreciation was abundantly accorded to him. In 1926 he retired from active duties, first to Madison, then to Pasadena. Of the many to whom he is now known, none can miss him as much as those few older and simpler friends who, through all these years, have loved him as a brother.

I shall close this account with a letter which Birge wrote to Burr Jones in connection with the fiftieth anniversary of the founding of Town and Gown.

Trout Lake, Wisconsin
August 5, 1928

~ My dear Jones:

So you want me to tell what is "most significant and important in
the history of the Club"?

Let me tell you, first, that your request is wrong in principle. The
Club, like the good woman of the classics, has no history. It has annals
and so you are a pro tempore annalist, not a historian, though you seem
to think yourself one. All the same your request, taken as it stands, is
easy to answer and here you have it.

First—important—no doubt its birth, at which you acted as parent,
or possibly Socrates might have said, as midwife. But I don't remember
that event and so it doesn't count. For me the most important event
in the past fifty years of the Club is the invitation which you gave me to
join it. I am perfectly clear on this point and you may consider it settled.

Second—significant—here also the situation is equally clear. There
has never been a woman at any of our meetings. If you want to play the
lawyer and quibble you may object that this is not an event but a non-
event. But surely the fact is of great significance that during a half century of universal and rapidly advancing feminization the Club has wholly
escaped the movement. This is all the more significant since the situation in other Clubs of our kind was different. You may recall that the x
Club of which Huxley was a member was flourishing when ours was
founded. It sometimes admitted women under the formula of invitation

$$x + yv's = ix.18.$$

But no such formula or fact appears on our records.

So much for your request, of which, as you see, I have a low opinion. For the specialty of the Club has been not to have a history of importance or significance. No one has ever succeeded in giving it a decided character. Take the intellectual side. We've all tried it, but with
what success? Owen no doubt represents the right wing of intellectual-
ism and he usually brings us some of his serious work. We listen with
interest but how far does he get with it? Any farther than either you or
I? Have the efforts of the three remaining founders* made us a "little
group of serious thinkers"? I trow not. We may do serious thinking when
necessity forces us but not on Saturday night or when we are together as
a Club.

Or look at the other side. Slichter obviously represents the left
wing of conviviality. But does he get far with that? I think he gets about
as far as did that uncorkable bottle of tequila (if you spell it that way,
I didn't taste it and so I can't spell it) which Overton once showed us.
If I may appeal to your profession, would not Counsellor Playdoll have

found the Club a rather dull substitute for the high jinks at Clorthugh's?

No, we have avoided that which makes history. We have met to cultivate and to enjoy the common sympathy which unites men of the same general type but who are working in fields widely remote from each other. We haven't tried to gain much knowledge or great inspiration from each other. Most frequently those who lead the talk go outside their profession or to a side aspect of it, rather than to its regular work. We tell of that which may well be common to us all or we bring in a book of similar appeal. We have said what we thought at the moment without much regard for consistency. We have tried to set free the "hidden soul of harmony" by the frank and spontaneous expression of the thought and feeling of the hour or the day, thus moving easily and naturally into that common life of ours to which the Club has contributed so much.

I don't see, therefore, that I can close better than by a reference (I think) to another member of your profession, Lord Melbourne. He said that what he liked most about the order of the Garter was that "It had none of this damned merit about it." And so the chief charm of the Club is that it has had neither importance nor significance; and you may qualify the nouns (provided Owen will let me call them nouns) with any adjectives (if there are such things) that you may have handy.

As ever, very faithfully yours,
E. A. Birge

* Charter members, rather. You are the founder.

There may be those who think that this discussion of the Club has been too extended; but I believe that, after his family and close relatives, no group of people so engaged Slichter's affection as the members of Town and Gown, and no other gatherings, again except for family, gave him such joy.

Besides the three clubs which were of major importance to Slichter —Town and Gown, Madison Literary Club, and the University Club —he belonged to numerous others. Two of these, the Madison Club and the Maple Bluff Country Club, owned buildings and provided eating and social or athletic facilities.

He also belonged to the Chaos Club, a group of midwest scientists who held most of their meetings in Chicago, but at least once a year visited the backcountry. The name is said to be a "spoofing" reference to the alleged stuffiness of the Cosmos Club of Washington. Slichter sometimes went to its meetings in Chicago and helped organize those held in Madison, the arrangements for which in the early days of his membership included carriages to take the visitors from the station to the campus and to deliver them back to the train. When he could not attend the meeting of October 28, 1905, he sent the following:

TO CHAOS

Fellows, ope the stein! To that Primal Mix
Of all that since has been, drink; and fix
Your long range guns upon the misty beach
Which skirts Time's utmost bound. Then reach
Yet far beyond, until we Chaos name—
That greatest First from which all lesser came—
That mixture both of Mind and Mass, the same
Joined tight in cosmic clasp, nor God
Nor matter then was named. A virgin clod
Of stellar jell, not yet inoculate
With germ to bring e'en Cosmos forth—I rate
Thee First!—First Cause—First Stuff! To thee I bid
All drink. . . . Then close upon the stein the lid.
'Tis time. So eat.

He belonged to the Six-O'Clock Club—dinner and short talks. He even had a life membership in the Palma Club, whose identity baffles the reference department of the University of Wisconsin Library.

He was a frequent guest at numerous dining clubs, especially at one called SPRC (Society for the Promotion of Research and Conversation), composed of scientists including some of his younger friends such as Max Mason, Warren Weaver, and Edwin B. Fred, and which still flourishes.

But his social life was not limited to clubs. Slichter was a marvelous dinner guest. I could just as well have started this paragraph with "Slichter as a dinner guest was a loud-voiced monologuist." Both statements are true; and that can only be the case if the monologues were superb. They were! When he was not holding forth, he could in his later years lapse into long silences sheltered by his deafness.

Once I was present when his right to the floor was challenged by Professor Caratheodory. Caratheodory was a Carl Schurz Professor visiting from the University of Munich, brought up in a diplomatic family, and a polyglot from infancy: "We spoke French and Greek in the family and learned English and German simultaneously. I do not remember when I could not talk four languages." One listened to him with interest, for there was rich content in what he said and the style was fascinating. One also listened to him perforce, for he would corner one with the lighted end of his evil-smelling cigar perilously close to one's nose.

On this occasion, with Slichter and Caratheodory both talking at once, the rest of us were enthralled and we would be rewarded by either of the possible outcomes. The contest was almost equal, but the cosmopolite with the cigar was finally overcome by the louder voice, the whiter hair, and the greater ability not to listen.

Slichter's sayings out of context would sound flat. Some have been worked in elsewhere and others, such as that on the dismissal of Glenn Frank, "A president who has not worked himself to death in ten years should be fired. Now we want one that will attract fine young men. My candidate is Wally Simpson" (the recently become Duchess of Windsor), will be lost with the memories of those who heard them.

Chapter 9 ~ *Essayist*

IN my judgment there has been no finer essayist on the faculty of the University of Wisconsin than Slichter, and few who reached the same level. Among the few were Aldo Leopold, who wrote in a somewhat different genre, and Birge.

We have already given, in Chapter 2, quotations from his high school and college writings. The range of topics was presumptuous: "The Abolition of the Monasteries," "Immigration," "Happiness," and "Civilization." He tended to be dogmatic and flamboyant. At times he assumed a tragic pose, "No full happiness in life—only in death." On many of his themes the instructor not only corrected spelling but, more important, tried to curb the youth's tendency to be carried away by the richness of a phrase independent of its meaning. Words such as "farcement" are questioned, the comment being "Obsolete— its meaning 'stuffing,' is possibly not what you wanted here."

"The Drama of Life" (January 1883, when Slichter was a sophomore) reveals the natural ham which helped make him a great teacher and the exuberance which—when married to restraint—produced a spirited style. He later became meticulous in his choice of words, gave up the tragic pose, and, God be praised! never lost his sense of humor. Two samples from this theme follow. We have retained his spelling.

~ *That all the world's a stage is the admission of every one, and, further, that it is an immoral stage, is the sad conviction of many who have seen much of it. In the dimness of this obscure, peaceful little nook we may not be able, or care, to detect the gluttonous farcement and pompous paint of our brother actors, but when, perchance, we view them before the brilliant footlights, we are no longer deceived by kings and fools*

and clowns, but discover in them only the one form of the hypocrite
and dissembler. . . .

We cannot hope to be acquainted with all the actions which are constantly occuring on a stage of such vast area. But History, a digest of all
the shifting of scenes and the accompaning confusion and commotion,
and what little we can see of the performance while we are actors,
teaches us the one great moral to be learned from this great drama of
life;—that is, after the last act, after the curtain has fallen for the last
time, those who have been the most akward with the mask and have
made poor hypocrites, these, I say, when all the paint and masks will
have been removed and all made alike and equal, will be the ones whose
conscience will uphold them. However much the pompous STARS may
have been laudied or envied, they will find small demand for their trade
when this theatre of tradegy and comedy lay in its predestined ashes.

After the faculty at Northwestern, no other group drew from him as
many papers (nine) as the Madison Literary Club. I list the dates
and titles of these below:

October 13, 1902, "Archimedes";
October 8, 1906, "The Desert";
June 12, 1911, "Industrialism";
November 13, 1916, "The Scientific School at Alexandria";
April 10, 1922, "The Royal Philosophers";
October 8, 1928, "The Club";
June 13, 1932, "Polymaths: Technicians, Specialists and Gen-
iuses";
January 9, 1939, "The Sideshows of Science";
January 10, 1944, "The Society of the Dilettanti."

His care in writing such papers is shown in his letter to me of November 15, 1938.

~ My dear Ingraham
I am asking a great favor of you. I have written (not at my own
suggestion!) a paper for the Literary Club. I can improve the diction
but I must have an independent judgement on its structure. To give human interest I introduce Rutherford. This seems to give me trouble in
producing balance. Should I approach the problem of human interest
in another way? I would like to have the paper as good as I can make it,
as it will be my last. It will be of great value to me if you can look it
over and give me your judgement.

Sincerely,
Slichter

I believe the first meeting of the Club which I attended was on April 10, 1922, upon Slichter's invitation.

In 1938 Slichter published *Science in a Tavern* (University of Wisconsin Press), a collection of ten delightful essays. There is no better way of getting the flavor, not only of Slichter's style but also of his personality, than by reading this little volume—by all means including the preface with its final paragraph: "I dedicate this little volume to the four sons of Mary Louise Slichter; it is these five who, with my former students, have educated me from a crude beginning to a most happy maturity." Four of the seven "Mad Lit" papers, which had been delivered before publication of this book, were included—namely, those of 1911, 1922, 1928, and 1932. In the second edition, which has also been printed in paperback, "The Sideshows of Science" was added. The other essays consist of "The New Philosophy," 1922, and "Heaven's Highway," 1936, both addresses given at the initiation ceremonies of Phi Beta Kappa at the University of Wisconsin; the second was at my request. (When Felix Frankfurter, not yet a Justice of the Supreme Court, received from a friend a copy of the first of these addresses, he wrote to Slichter praising it, sharing its worries, and asking for four copies.) I believe that Birge gave three such addresses; two other persons, as well as Slichter, have given two. "The Principia and the Modern Age" was presented to the Mathematical Association of America in commemoration of the 250th anniversary of the publication of Newton's *Principia*. "Science and Reality," "Science and Authority," and my favorite—"The Self-Training of a Teacher," which is extensively quoted in Chapter 3—complete the volume.

At the time these essays were published, the University Press, as well as other portions of the University, was still feeling the effect of the depression. Publication was planned for 1936 but did not take place until 1938 when Professor Harry Steenbock generously offered to subsidize it by buying for his own and Slichter's use the major portion of an edition of 500 copies.

The books received as gifts brought a flood of appreciative notes, not only for the essays but for Slichter's influence on many persons, particularly his former students. Some of the recollections they included with their thanks have been used throughout this book. Especially appreciative notes came from Ambassador Joseph Davies and from Professor Howard Mumford Jones, both of whom had been students of Slichter's.

Originally it was planned to use the title of one of the essays, "Heaven's Highway," for the book. In a letter to me of June 1, 1936, the Dean wrote: "I wish I could have your help in inventing a suitable title. The truthful one, of course, would be: 'Saved from the Waste-basket,' but that might be too truthful." I remember distinctly urging him to use "Science in a Tavern," a title which I believe I suggested, but that belief may come from the familiar trick of memory of recollecting what one wishes. In any case, it is my wife whose pen and ink drawing embellishes the title page.

Besides the essays listed above, there should be mentioned Slichter's paper on the administration of President Chadbourne published in the *Wisconsin Alumnus* of July 1940, and one, a gem, "Galileo," which was reprinted as a pamphlet from the *American Scientist*.

It may seem superfluous to describe Slichter's style when one can so readily sample it. But then critics would be on relief!

First of all, it should be noted that almost all of his mature non-professional writings consisted of material which was read by him on some occasion. Basically, they are a record of oral communication and although effective when printed, were even more so when read by Slichter in his resonant voice and with every evidence of delight in what he was doing. He never let one drowse. Almost always in the first minute or two, sometimes with the first sentence, there was a happy surprise that alerted one to the humor and the zest of the speaker. The subsequent series of epigrams, unusual twists, and magnificently stated ideas came at a tempo that never permitted one to relax, let alone to become bored. The first essay in *Science in a Tavern* began with:

~ Not all the heavy eaters and hard drinkers of past centuries have been poets and dramatists, or literary fellows. I propose to show that the men who led scientific progress in England during the seventeenth and eighteenth centuries were men who could and did eat and drink with almost as much gusto as did Shakespeare, Ben Jonson, Marlowe, Donne, Dekker, Dr. Johnson, Boswell, and other luminaries of literature and the arts.

It was a delight in the dreary waste of listening to papers at a mathematical meeting to hear one which started:

~ The beginning of a unified conception of the universe was not made by mathematicians. It was the discovery of poets. That is to say, the first approach to nature was esthetic, not scientific. Over two thousand years ago Solomon is reported to have acknowledged his God in these words: "He hath given me the certain knowledge of the things that are, namely to know how the world was made, and the operation of the elements; the beginning, ending, and the midst of the times; the alterations of the *turning of the sun, and the change of seasons; the circuit of the years and the position of the stars: . . . And all such things as are either secret or manifest, them I know."*

These are strong words from Solomon. His enthusiasm seems to have outrun his modesty. Only poets can indulge in such overweening self-confidence. They call it inspiration. Sometimes I think it is as sound an approach to reality as any other.

One might conclude from the words attributed to Solomon that he and not Newton had written the Principia.

Slichter loved to illuminate the general by the tidbits of history and float the particular upon the river of human change.

The first of these traits is well illustrated by his joyous paper on "The Club," in particular, the literary club founded by Joshua Reynolds, where neither the stimulus of wine nor of wit was absent. One chortles to find that in 1851 Macaulay set down in order the six great poets as Shakespeare, Homer, Dante, Aeschylus, Milton and Sophocles; and that Carlisle tried without success to somehow let Virgil muscle in on the group. (I would reorder them as Dante, Shakespeare, Sophocles, Homer, Milton, and sixth, if included at all, Aeschylus.)

And what a pleasure to find a man of seventy-eight years begin a paper on "Galileo" with:

~ *Galileo was born in 1564, the year in which Michelangelo died. Galileo died in 1642, the year in which Newton was born. Thus the history of 250 years of human greatness can be written with the use of only three words—Michelangelo, Galileo, Newton. All other words contribute mere details, padding, or dilatation. Michelangelo had contemporaries, meaning thereby those whose life span substantially overlapped his own. Some of these were Perugino, Signorelli, Leonardo, Correggio, Titian, Raphael, Verrocchio, Cellini. Galileo had contemporaries— Stevinus, Gilbert, Kepler, Descartes. Newton had contemporaries— Huyghens, Leibnitz, Fermat, Pascal. What does this enlarged vocabulary signify? Do these words mean that the age was changing, that a new curiosity had been aroused, that the creative imagination had been transferred from art to science? Yes, these words mean all of that, but of course they do not mean that this change in the pattern of events was a novelty of human history. It does not mean that a shift in interest is unique or rare. It is stability that is rare. Change, new values, new philosophies, new interpretations are the normal expectation, just as the hats, hose, and doublets of one year are expected to be out-styled by the raiment of the next year. Life seems to move in reversals, discontinuities, and contradictions, rather than in smooth continuous progress.*

Seldom do we find an era of long-continued stability or epochs of changeless beliefs or a long period of abundance, peace, and happiness. Michelangelo stood at the culmination of an amazing period of pictorial and plastic art. Newton headed an era of the full power of science. Galileo was the link, the new prophet or the John-the-Baptist of the new philosophy who joined together the two dispensations. The words "Michelangelo, Galileo, Newton" make up the vocabulary of a transform in human history.

The pleasure is maintained to the conclusion:

~ Thomas Aquinas, you are not dead! Listen to the moderns: A whole is more than the sum of its parts! That is what you said! "The form of things is supreme over the mere substance of things." They even tell us that the form of space is supreme over the course of the stars!

Roger Bacon, you did not die! The nineteenth century was your century.

Aristotle, you are immortal. Your claim of purpose in the world cannot be repudiated.

Kepler, you live as the example of the well-mannered and tactful exponent of new ideas, smoothing the way for general acceptance.

Galileo, you also live; for you are the model of the bond-breaker, that denies control of science by outsiders and frees it from the rule of authority.

But more than all, there persists the perpetual precept: "There is Room for Difference of Opinion!"

Ages slowly recede and we condemn them for what we ourselves are now doing. The Middle Ages burned witches because they were violating the teachings of religion. We in this age burn or destroy Poles, Norwegians, Czechs, Dutchmen because of a quibble over political minutiae. We burn books and works of art and we denounce as criminals a host of scientists, poets, and religious leaders whose teachings do not conform to the fiats of a political bureau. We condemn Jews because they are Jews and in this country we condemn Negroes to economic and educational "jimcrowism" because of the color of their skin. We should publish our criticism of the Middle Ages in humility and with muted voice, as we turn to the confessional with our guilty hearts.

Pascal! You are not dead! The spirit of evil still has miraculous power and it takes all the force of the world to overcome it!

In the brief moment in the morning between awaking and becoming fully awake, in that fleeting dreamland where the mind is really clear and the spirit is fresh, I am seized with an ecstasy of revenge and hatred and I demand that the Devil and all his works be purged from the world by fire! In that shadowland, in that ghostland, I live for a moment with Pascal, amid the blessed realities of the Middle Centuries.

In the middle of the paper we have:

~ Up to this point I have tried to recall to you the environment within which Galileo spent his life and to describe the atmosphere breathed by his learned contemporaries. As I am neither a philosopher nor a historian, the picture I present must be imperfect. There is one aspect, however, that I am quite sure is correct; namely, that among all the diversities, dissensions, and disputes of that day, and for similar diversities of all other years, ancient or modern, there has always been room for difference of opinion. In matters deeply affecting human society, we are sure of our ground if we put at the top of our mast of philosophy the slogan: "There is room for difference of opinion." There is no world matter of consequence that does not require tolerance. There is nothing that is not brought nearer to perfection by an unbiased approach and fair discussion. It is this attitude that I have tried to build up for you in this essay; it is in that spirit that the age of Galileo must be understood.

We have his own discussion of this paper. "Galileo" was delivered to the University of Wisconsin Chapter of Sigma Xi on November 17, 1942. It was planned that it should also be presented at a mathematics meeting in New York on December 30, but the scientific program was canceled due to the war travel stringency. Between these two dates there was an exchange of letters with Professor Max Otto of the Department of Philosophy at the University of Wisconsin, who had heard and greatly liked the paper and wrote to him, saying of the address: "It seems to be the result of a certain temperament, a certain playfulness of mind, a certain meditative response to life combined with the experiences which the years bring to a person gifted in this way." The following paragraphs are contained in Slichter's reply.

~ As you probably noted, the plan of the paper is simple: First paint the picture of the age of Galileo. Next avoid too much attention to the actual facts discovered by Galileo. Next build the principle climax on pathos—the Cordelia and Lear pathos, where he forsakes the security of Venetia for the uncertainty of and insecurity of Tuscany because of his deep affection for his daughter. That was true tragedy which I am sure I did not exaggerate. I then built the secondary climax on the partial revival of scholasticism in the present day (led by Einstein) and at the end I called the roll of the great actors in the drama—Aristotle, Thomas Aquinas, Roger Bacon, Kepler, Galileo, Pascal, to bear witness to the modern day.

All of this is imperfect and more so because it is a 24 page paper and not a book. When in college, Herbert Spencer was the man. Every word of his First Principles was gospel to our green group of college enthusiasts. I have spent all my life hoping that soon I would find Science exerting its redeeming power over human behavior and human conduct. It has been the major disappointment of my life that, as yet, I find no (or little) influence on human behavior coming from science. Before this war, there seemed to be hope that science was actually raising the standard of living, which might eventually result in an improvement in human behavior. But it takes more than good food, good houses, good autos, good plumbing to make a man. I have thought that perhaps psychology would soon reach an adequacy and a power so that an impartial influence might be brought to bear upon human behavior. The power, it seems to me must come from adequate knowledge of fully developed psychology (of fifty years hence?), recognized by the public and educators alike, as sufficient to control, in part at least, the evil that is in man. In the meantime, in my despair, I join Pascal and try to meet the devil with the purge of fire. . . . The great job, as ever, is the betterment of human conduct.

Again I thank you for your appreciative letter. Even an old man needs encouragement.

I have quoted at length from, and about, this essay, for in it with no loss of vigor we find the mellowness of Slichter's age—a mellowness which was not the characteristic of his youth. First vigor, then vigor plus wisdom, and finally vigor plus wisdom plus mellowness; yet a mellowness which permitted fire, indignation, and perhaps not only the appreciation of the poet but the willingness to be the poet.

Chapter 10 ☙ *The Bureau Drawer*

Nature and Title of This Chapter

THIS chapter is an admission of defeat. Slichter's energy found expression in so many ways that, struggle as best I could, I did not succeed in compressing the story under a few neat headings for, as he wrote in concluding "The Side Shows of Science," "Something has been left out!" There are many things I do not wish to leave out, but know not where to place. Some of them are gathered together here.

I do not really know how Slichter filed. On occasion he was successful. His secretaries, when he was dean, usually were. Frequently, however, he had folders which contained not only much relevant to their labels, but also other material which now is more effectively lost than if never filed. Often he kept what he wanted in a fine hodge-podge.

Occasionally he made an effort at order. In 1905 he created a file labeled "Bills (unpaid)." In it there are still a dozen items, dating from 1905 to 1907, the largest being for $14.65. I presume these bills were paid, but I have my doubts about one for books allegedly purchased in the late 1880s from a local bookstore which dunned him for at least four years.

I have a mental picture of him sitting between a wastebasket and an open bureau drawer. What disgusted him, he threw away; what he liked—letters of praise including those from Justice Frankfurter and Justice Brandeis, seed catalogs, receipts for oatmeal, family photographs, and a lock of Sumner's hair when he was two years, four months old—went into the drawer. It is from this creation of my imagination that I have taken the title of this chapter. I originally designated it "Olla Podrida."

Health

Although bodily difficulties seldom slowed Slichter's amazing activity, he suffered from three serious physical ailments. Starting sometime in his thirties, he became increasingly deaf; he was for many years on the edge of diabetes, with a low sugar tolerance; and in later life he often had severe attacks of sciatica. The last of these sometimes incapacitated him for short periods—for example, he did not attend the dinner given in honor of Professor Van Vleck, Professor Skinner, and himself, when the three of them, within a year, reached seventy. (He did, however, read what Warren Weaver was to say and wrote a reply which was read by President Frank.) During the subsequent fall he reported, "Began this summer as a cripple from sciatica but recovered by a miracle." He avoided too much sugar, although it was said that one could gauge how much he had taken for breakfast from the shortness of his temper.

The deafness seems to have troubled him more than either his back or his diabetic condition. Sometime in the 1890s he seems to have had abscesses in one of his ears. The prescription would indicate that these were accompanied by considerable pain. By around 1900, perhaps but not necessarily caused by the abscesses, he had lost the hearing in his right ear, but wrote later that this "caused very little inconvenience." But in 1930 he began rapidly to lose the hearing in his left ear. Various treatments, including the extraction of teeth and heat therapy, did little good. His experience with hearing aids was exasperating, for at first he received no instructions with a defective instrument with one part missing—faults which he discovered as each previous one was remedied. Finally, however, he was able to write to his son Sumner: "I have a new hearing device (as I think I told you) and it seems to be a great help. Tonight I shall try it out on Helen Hays at the Parkway. Wednesday I will try it out at a formal dinner at the Bucks and Saturday at my own Town and Gown at the College Club."

It was his partial deafness—when, after he had paid little attention to the discussion in an administrative committee meeting, he arose in wrath at the suggestion of cutting an item in the budget of the Graduate School—which led Dean Sellery to accuse him of having "damned selective hearing."

Do-It-Yourselfer

Slichter was a do-it-yourself man. He did experimental as well as theoretical work on the flow of water through sand. He helped in the building of Sladshack, his cottage on Lake Mendota. He gardened. He washed the car, at least after his sons were no longer living at home, for in a letter of uncertain date he wrote to Louis: "Washed the car in ice water Monday. The side yard was frozen hard and I ran the car in and put the hose out the window. For two hours I divided the water between the car and myself. The car got clean and I got dirty. Such is life." Another washing included cleaning the carburetor. He made a self-feeder for his hens which would keep them, at least so he expected, from scattering seed on the ground; a baler for waste paper; and designed an easy way of putting their pier up each year—this is still in use.

To his son Sumner he reported with joy his manual labors:

Woodsman: "I went to the cottage and cut down an oak stump that had been overhanging the bank for years and then wedged it in pieces for the big fireplace."

Plumber: "Tomorrow . . . I will take the varnish off of a little table and also start rearranging the waste pipe from our kitchen sink."

Electrician: "I have done a good electrical job—as a surprise to your Mother but she has not yet seen it to approve."

Painter: "The sills . . . must also be painted. I have enough of the brown paint to do it. I'll do it at once. . . ."

Packer: (Not quite as successful.) "When you write tell us how many plates were broken—I packed them and thought I knew how to do it. Perhaps it was foolish of me not to take them to a professional packer."

Carpenter: "I have made new cellar stairs for the back cellar door— quite a job as I had to start with very hard 2 × 12 yellow pine plank. I have several days work of other sorts in the cellar, all of which I enjoy."

Cabinetmaker: "I am making a table for the kitchen in Madison. I am using a cypress board that I have had for some time about 24″ wide for the top."

As witness of his skill as a cabinetmaker, there is a letter of Allen's to his parents written on Sunday evening but otherwise undated:

~ *Dad's most excellent piece of cabinet work—my record album cabinet—now graces our living room, next to our Capehart, and besides being most useful and ornamental is one of my proudest possessions. Dotty and Marjorie brought it back with them Friday—and I want to thank you again for such a grand gift—one that will always be especially cherished because you made it yourself, and did such a swell job of it. How often do you find cabinetmaking and mathematics developed in*

*one individual to such a point of perfection. I could also add to that
literary and culinary achievement!*

I wish we had more details of the last-mentioned accomplishment,
which we still assume was better left to Mrs. Slichter.

Apparently there was one lacuna in his manual skills, for in the
minutes of Town and Gown for April 9, 1904, Judge Winslow recorded: "The host well remembers when Slichter attempted to carve
ducks. It was an exciting time for those who sat near."

Both his workshop and his garden lured him to Sladshack.

He appreciated good work on his own part and on that of others.
His son Louis reports that such work gave him occasion for his only
frank outbursts of self-esteem. (Louis must have forgotten his joyful
look when he produced a really successful witticism.) He would say
when his craftsmanship turned out well, "It takes brains to do that
kind of a job." Louis adds, "By contrast or by implication, of course,
it took none to do anything connected with the University." Slichter
met his match in a truly excellent carpenter, Mr. Marsh of Waunakee,
who would not let his wages be raised above fifty cents an hour, in-
sisting no carpenter was worth more and that he couldn't take the care
to do the job right if his time was costing so much.

Two weeks after the attack on Pearl Harbor, Slichter reported, "I
am busy making and repairing things. Such work eases my mind."

Robberies and Thefts

Slichter seems to have been prone to losing property through theft
or robbery. The most exciting event of this nature occurred on April
20, 1925, in Chicago, when he was boarding a Pullman car bound for
New York on the New York Central Railroad. His description of this
in a letter to the Pullman Company, written ten days later, states:

*~ I entered car 426 after having given my grip to a porter. As I went
through the narrow aisle leading into the main body of the car, I was
followed by one party and as I was about to enter the main body of the
car, two men were in front of me. As the porter returned he asked them
for their reservations to which they replied they were seeking a gentleman giving a name something like Mr. Ferguson. They almost immediately rushed me near the entrance of the main body of the car sticking
their elbows in my ribs and holding me in this grip. They almost immediately released me and upon proceeding a few steps, it dawned
upon me that I had been robbed, and feeling for my purse I found that
that was a fact. I immediately gave chase, but descending the steps,*

*they were not in sight, showing that they had passed into car 404. I im-
mediately notified your representatives and representatives of the railway company, but the parties evidently had disappeared on the yard
side of the train.*

He was explicit as to how much he lost.

~ *I left Madison with eight $50 bills and $9 in small paper. I used
$40.38 at the Madison office, paying this with a $50 bill. $9 in change
was returned to the pocketbook. At the Chicago office I paid $16.20 or
about that amount for pullman fare with a $50 bill, returning to my
pocketbook the paper change amounting to $33.00. $1.00 in paper was
used at the Northwestern station in the morning. This would leave exactly $350 in paper in the pocketbook.*

He made a claim for that amount and finally threatened suit without
much enthusiasm on the part of his lawyer. I have a vague memory
of being told that the matter was settled out of court. This robbery
became a cause célèbre for Town and Gown.

In January 1905, Mrs. Slichter's jewelry box, containing a gold neck-
lace and a pearl pendant, was stolen from their hotel room in Washington, D.C.

On a Saturday afternoon—year not stated but probably after 1937
since it is on hotel stationery collected in Florence—Sumner wrote a
letter to his parents chiding them for leaving the door unlocked at 636
Frances Street which had resulted in a sneak thief making off with
some property.

Hens and eggs were lost at Sladshack; more seriously, on another
occasion a lathe was taken; and, yet again, a fine set of wrenches, plus
at least one egg and an old pair of trousers. We do not know what was
lost in the house (Sladshack), but Slichter finally installed a "tear gas
grenade" which would be tripped by a wire and the "property protected at least until plenty of time has been taken to completely air
out the premises."

I have an unholy suspicion that a man with a wife who did not approve of losing things could readily believe that minor mislaid items
had been stolen. We know he accused boys of taking trees from "Governor's Island" (in Lake Mendota) only to find that they had permission.

Two Versions of One Story

We do not know for what occasion he found appropriate the first
of the two versions of the account of "Spot(s)." The second was used

at a breakfast arranged by George Haight for a group of alumni. Both
stories are undated.

First Version.

BILL WILLIAMS: HIS DOG SPOTS
Charles S. Slichter

~ *Bill Williams was an Oklahoma wildcatter. He called his dog Spots
not because it had any spots on it, for as a matter of fact it was a long
legged pet without any markings at all—in truth the dog was of a uniform dull color exactly the shade of Oklahoma crude oil. He was called
Spots because of his occult ability of spotting Oklahoma oil pools. Bill
Williams soon noticed that when Spots found a convenient tree or post,
he was in the habit of raising one of his long hind legs and pointing it
directly to a well known oil pool. Bill wondered at this and soon concluded that Spots possessed occult powers—"mind reading" he called it,
but in reality it was not "mind reading" but "oil reading." Bill Williams
was determined to try his dog out in virgin country where no oil was
known to exist. When Spots pointed a hind leg in a certain direction,
having found a convenient tree or post, Bill followed that direction and
put down a wildcat well. He nearly always struck oil, but he sometimes
missed it because he found it difficult to follow the down slope or
declination of that uplifted hind leg. He then hit upon the scheme of
taking direction bearings from two different locations selected by Spots.
Where these directions intersected Bill put down his discovery well
and was lucky to find that he never missed at all—not once. All of this
made Bill Williams a very prosperous man and he sort of retired to an
Oklahoma town called Jackie. In cowboy days this town had been called
Jackpot, but the railroad and the U.S. Post Office Department had sof-
tened the name to Jackie. Near Jackie was a pipe line pumping station
equipped with a round tank to take care of the slack in pipe flow. This
tank was surrounded, as usual, by a moat and a circular fence supported
on strong wooden posts. In order to demonstrate to friends and visitors
the "mind reading" qualities of his dog, Bill would walk Spots around
the circular fence in clockwise direction. Spots would always raise his
right hind leg (at a suitable post) and point to the oil tank. When he
led him around in the opposite direction, sure enough he always lifted
his left leg toward the crude oil. This was the demonstration. One of the
pumpmen in charge of a shift at that pumping station was Mike Hogan,
a man said to be of Irish descent. Hogan had been a pipe fitter for the
Dome Oil Co., but when the work got too heavy for him, he had been
given the lighter job of pumpman at the Jackie Station. He watched*

the demonstrations of that dog and in consequence he soon became excited and nervous. In fact he could hardly sleep. It was just like the time he visited his brother in Colorado—there the air was so free from the pungent fumes of Oklahoma crude that he nearly suffocated—he had to hurry back to Oklahoma where the reek of crude could again put his lungs in working condition. Hogan's nervous troubles got worse until a branch line was brought into Jackie requiring that the old tank be replaced by a larger tank. Mike persuaded the engineer to locate the new tank on the opposite side of the line. When the old tank was demolished, Mike took all his savings—$8000—and with the aid of a speculative driller, put down a well in the exact center of the circle of the old tank location. At 4000 feet they brought in one of the best producers in Oklahoma. Then Mike had to explain. He had noticed that the hind leg of Spots had never pointed to the crude oil in the tank but to a point far below it. The fact is that Spots is not sensitive to crude oil in a tank but only to crude oil in underground pools. By logical reasoning Mike had made himself a millionaire.

If any one wishes an autograph of either Bill Williams or Mike Hogan, write to the Dome Oil Company at Oklahoma City. Both men are now on the Board of Directors of that prosperous company. If one of them should happen to put his autograph for you on a bank check— which he probably will not do—there is no reason at all why it should be for a small sum of money.

Second Version.
THE STORY OF SPOT
~ Spot was the name of Mr. Williams' dog. Now the peculiar thing about Spot was the fact that he did not have a spot on him. He was named Spot because he had the unique ability to pick out famous people—celebrities. He could spot such persons wherever and whenever he found them.

When he would go out walking with Williams on the streets, Spot would always pick out the distinguished persons. He never failed, e.g., a judge, a preacher, a famous business man or even a movie actress. Here is the way he picked out the famous people. He would stand just like a bird dog does for a covey of birds—his tail and front leg up and nose pointing toward the famous man.

Now in all of his life Spot never was known to stand or spot a Dean or a President of a University.

Once upon a time Williams took Spot with him to a reunion of his class—an occasion just like this and when Spot walked in and saw the various members of the class he looked all around and what do you

*think happened? He saw so many famous persons that he did not know
what to do. He turned round and round pointing in all directions.*

Charities

I find little record of Slichter's conventional charities. He supported
Grace Church generously and the family had a pew there. Occasionally
he gave to the Salvation Army and the Volunteers of America, but
chiefly he liked to make gifts into which he could enter personally and
which would lead to even more substantial giving by others. I have a
vivid and, I judge from the records, an only slightly inaccurate recollection of a statement he made to me. "I visited the University of the
South. That's a good institution. They make the freshmen take Greek.
I gave them a good book and I told Warren Weaver and they are getting some money. And there's Northland College. It hasn't much
money but it's a good small college. Those Swedes work hard. I gave
them a good book and I told George Haight about it and they're going
to get some money. Now, Ingraham, I have another good book. Do
you know another good small college?" I wish I could claim to be responsible for the sequel, for the other "good small college" turned out
to be Washington and Lee. My hunch is that E. B. Fred pushed Washington and Lee, the shrine of Lee and his horse, Traveler; but moreover it was also a favorite college of Slichter's son Allen.

According to the letter of thanks written by President Guerry in
1942, Slichter gave to the University of the South forty-four volumes
of facsimiles of the quartos of Shakespeare's plays, and Sydney Lee's
facsimile of a first folio of Shakespeare, plus supporting bibliographical
material. He had previously given to that university his files of the
Bulletin of The American Mathematical Society.

In March of 1943 Northland received a copy of the Lee facsimile,
and sometime that same spring he made a similar gift to Washington
and Lee. He appreciated both Shakespeare and applied science and
wished to help assure that each would benefit society.

We do not know how his financial prognosis worked out; but in
connection with his interest in Northland College, Mrs. Slichter's
niece Grace Merrill Hodgkins, who lived in Ashland, wrote on August
18, 1943, "Thought you might be interested in knowing that the seed
you sowed last winter when Walter and I walked in and you said you
were thinking about the future of Northland College, has commenced
to grow." She then described the successful start of a money-raising
campaign.

There were gifts of a promotional nature to advance the scientific activities of the University of Wisconsin. Two of these—the first, to start a fund for Max Mason's work in submarine detection; and the second, an advance towards financing the first foreign patents of the Steenbock process—are described in Chapter 6.

The more purely humanitarian interests also had appeal to him. Warm letters of thanks are our only record of one such charity. These letters he received from various priests in Austria for the condensed milk he sent in 1920 for children in Vienna, Linz, Salzburg and Innsbruck. I quote one in full:

> *Salzburg, April the 14th, 1920*
> ~ *Yesterday we got the first package containing evaporated milk for children and babies of our city.*
> *In the name of poor and hungry children I thank you very much. We gave the milk to a child-house under the direction of sisters.*
> *Children and sisters will pray for you.*
> *Anton Keil,*
> *Regens of the seminary in Salzburg*

This reminds one of his enthusiasm for the cod-liver oil, fortified by the Steenbock process, which enhanced the health of the very young, and of the promptness with which, forgetting the animosities of World War I, he worked as dean to reestablish the Carl Schurz Professorship.

Respectful Iconoclast

Slichter was reverent but loved to shock people, a trait already mentioned in Chapter 5. He poked fun at those who took themselves seriously—for instances, (1) Phi Beta Kappa, (2) temperance workers, (3) overlong rituals, and (4) Harvard—and he punctured the inflated. As evidence:

(1) He started his address, "The New Philosophy," with "Phi Beta Kappa is a philosophical society, but it is easier to prove this by interpreting the symbol S.P. (Societas Philosophiae) on the reverse of the badge than by interrogating the individual members."

(2) In Chapter 7 we have quoted his letter to the Anti-Saloon League.

(3) While he was attending a Roman Catholic wedding of, I believe, one of his secretaries, his stage whisper to his wife—when the service had been prolonged for some while—was distinctly heard by those around, "Lou, we must have been living in sin for forty years." It was probably after this wedding that he told the bride, "I married an Irishman too. You'll make it work."

(4) When taking a bus to Harvard Square in Cambridge, as he approached his destination he said to the driver, loud enough to be clearly overheard, "What are all those buildings over there?" "Why, that is Harvard University." Then as he stepped out of the bus, rumpling his white hair in a characteristic gesture of self-satisfaction, he called back, "That's funny, I've lived here thirty years and never heard of it."

Even if he enjoyed ridiculing the pompous, he had the greatest respect for the visions of youth, for self-reliance, for brains, for hard work, and for friendship. His attachment to his family was complete. His approach to reality was that of the mystic as well as the scientist. He stood in awe before the grandeur of nature; and though he was fascinated by the geniuses who had partially discovered some of her secrets, he had little faith that all of them would ever be revealed.

President Fred likes to quote Slichter's dictum, "It is always safe to doubt many things written down in the sacred scriptures of science."

On March 13, 1940, Slichter wrote to his son Sumner:

~ I cannot place the mystic on a lower level than the scientist. Both report only a small part of the truth, but both parts seem to be needed. The ancient astronomers viewed the fixed stars and memorized them in constellations and worked out the motions of the wandering stars and of the moon and sun and told the seasons. But the psalmist noted these same stars and from them read a message that exalted men to a new courage of the spirit and enriched them with visions of far-off purposes and purified them with the awe of its beauty.

A Letter from Turner

I have been unable to discover with certainty the occasion for the following letter to Slichter from Frederick Jackson Turner. On February 3, 1923, Slichter had given in Chicago a Founder's Day talk of which I have found neither an outline nor a copy. It is possible he took this occasion to blast whatever was his current favorite object of scorn, and that this letter is a commentary upon the talk. In any case, with just a little aid from the imagination, Turner's reply can stand by itself.

7 Phillips Place
Cambridge 38, Mass
Feb. 23, 1923

~ Dear Slichter
        But I didn't mean to "demolish" any thing, and at close range I should "dig in" and adopt a "passive resistance" for "safety first." I

agree with you that the Universities need challenging—it's good for 'em!

The change is being made, both in the kind of instruction, the emphasis upon independent reading and student initiative and the correlation of his studies. Out here we suppose that Wisconsin, at least, is in touch with the farm and the shop as well as with the laboratory of the University.

I agree that field work in Economics and in other of the "humanities" is important. But I don't agree that it can take the place of the University. There seem to be plenty of rocks falling upon the University head anyway, and, as for the overalls, they may be desirable for that part of the economics students who are trying to ascertain what labor is thinking. This is worth while; but no economist will get far in leadership toward the solution or amelioration of problems who bases his action fundamentally upon the ideas, prejudices, and passions of the man in overalls, any more than he will by serving as the clerk in a bank, or the seller of dry goods at the counter of a store, or even by becoming labor union leader or railroad president. What is needed is disinterested and informed independent study of a way out, rather than saturation with the feelings of any of these classes. I do not think that careful and intelligent economists who study the labor question (Commons, for example) are unaware of what goes on in the back of the head of the laboring man, and I.W.W. ideas are known through their literature, etc. I can see that the student might come better to understand the feelings and the mentality of the I.W.W. working man by being hit on the head by rocks from the mines, but would he really be in a better position to deal wisely with the whole subject for spending half his time there? If you could put the capitalist and his sons down there they might have something knocked into their heads. But I question how much further on you would get the economist, per se. I haven't any brief for the economist, but I think you underestimate the extent of their reading of their own original sources, the organs of unions, trade papers, governmental and other reports, financial papers, bank reports, trade statistics, etc. etc., as well as their actual contacts with the men of affairs, labor leaders, farmers, and so on. By the time your boy or professor has lived long enough in his mine or his workshop, or on his farm, or in his counting house, or on his freight car, or on the docks as stevedore, or has polished up the handle of the big front door, and cooked the dinner,—and actively shared the life of the many classes and sorts of workers, he may realize that it is a darned hard life and may qualify for I.W.W. membership; but will he really be a wiser economist for it all?—and when will he find the time to do anything but share the hardships of the world? Perhaps such an experience would make the average voter more sympathetic with the difficulties, hardships, and ways of thinking of these different groups. But wouldn't he have to turn to the scientific student, reasoner, umpire, perhaps between contending groups after all? Wouldn't he need the training in handling evidence from diverse sources, in analysis of the conditions, in fair mindedness and the reasoning power which a proper University training should furnish?

If not why not? But don't try to answer this disquisition. Wait until we can talk about it over the fence some time.

Seriously I liked your address very much. But what you really want is a University more in contact with the life and thought of the current world of affairs—and that is possible without substituting the mine for the college, or turning the course over to the man in overalls, —as I see it. But I confess that my overalls are used for fishing trips— though I wouldn't have missed my training as printers' devil and typo for a good deal.

Yours cordially
Frederick J. Turner

Gathered from Others

This section includes a few items which should be rescued from oblivion. They are from letters which Slichter received.

Justice John B. Winslow relayed to Slichter a newspaper report of a speech by Chancellor Day of Syracuse, "Chancellor Day has broken his long and extremely popular silence by making a speech."

A subtitle of a teacher's agency was, "The National Organization of Brain-Brokers."

From grandson Bill, August 3, 1933, "You can get in a lot more trouble by failing to salute one officer than you can by flunking half a dozen written quizzes."

Tributes

The following tributes to Dean Slichter are included here in spite of their saying much that appears elsewhere in this book, sometimes in almost identical words. The first is a resolution of the Wisconsin Alumni Research Foundation. (I do not know by what alchemy it made him into a dean of the College of Letters and Science.) The second is the resolution of the Faculty of the University of Wisconsin adopted December 2, 1946. It was drafted by a group of his friends who knew of his great services to the University. They believed he would have liked it.

RESOLUTION BY

WISCONSIN ALUMNI RESEARCH FOUNDATION

October 8, 1946

~ CHARLES S. SLICHTER *departed this life on October 4, 1946. He served mankind well. For his many accomplishments he was widely known both at home and abroad. He ably served the University of Wisconsin as Professor, and as Dean of the College of Letters and Science and of the Graduate School. He aided thousands of students as an educator and an advisor in their manifold undertakings. They have ever taken pride in having known him—he took pride in their advances in the University and in the after years. His genius in fields of mathematics, in science and in literature was evidenced throughout his long life.*

His mind was keen, his power of analysis superb and his knowledge wide. All of this was balanced by common sense and a levening humor.

He was a teacher who could teach. He did teach. He was able and willing in his leadership.

He was tolerant but with strengths unimpaired, a loyal friend, a delightful companion and a charming broad gauged Christian gentleman.

The resolution, which here follows, is directed to an expression of gratitude to him for his great help in the founding of the Wisconsin Alumni Research Foundation. He was vastly interested in the scientific discoveries of great human helpfulness made by Professor Harry Steenbock. With him, Dean Slichter saw the great power for good inherent in those discoveries, and encouraged Dr. Steenbock in his plan to follow practical methods in employing that power. He saw this Foundation's birth and aided in its undertakings to contribute the results of scientific research to mankind and in training men and women for research in Natural Science fields in order to best effect the advance of knowledge therein.

THEREFORE, BE IT RESOLVED, *that this Board, wholly composed of Alumni of the University of Wisconsin, is grateful to* CHARLES S. SLICHTER *for his great helpfulness to them and to the Wisconsin Alumni Research Foundation in its work of over twenty years, that this expression be spread upon the Foundation's records, and that copies thereof be sent by the Secretary to the surviving members of his family.*

RESOLUTIONS OF THE FACULTY OF THE UNIVERSITY OF WISCONSIN
ON THE DEATH OF DEAN CHARLES S. SLICHTER
Charles Sumner Slichter—1864–1946

~ Charles Sumner Slichter was born at St. Paul, Minnesota, April 16, 1864, and died at Madison, Wisconsin, October 4, 1946. He graduated from Northwestern University with the degree of Bachelor of Science in 1885, and received the degree of Master of Science from the same institution in 1886. Immediately thereafter, he came to the University of Wisconsin as an instructor and remained here until his death. He was instructor of Mathematics from 1886 to 1889, assistant Professor of Mathematics 1889–1892 and Professor of Mathematics 1892–1934. He became Dean of the Graduate School in 1920 and retired as Professor and Dean Emeritus in 1934. During the period from 1934 to 1939 he served as Research Advisor in the Graduate School.

In 1916, he received the honorary degree of Doctor of Science from Northwestern University. Starting in 1898, for many years he was consulting engineer to the U.S. Geological Survey.

In 1890 he married Mary Louise Byrne who with their four sons Sumner, Louis, Allen and Donald survives him.

Such a record might seem to be that of an uneventful life. This is not the case. Dean Slichter had a rich and colorful personality and each day was to him a vivid experience and a lively event for those with whom he was.

Dean Slichter's contributions to the University were manifold but were chiefly as a teacher, a scientist and an administrator.

His success as a teacher was not limited by the degree to which he inspired men to become mathematicians. He used mathematics as a vehicle for a liberal education, as a means of enriching the lives of men. Few teachers have given the student more insight into the relation of Mathematics to the whole range of science or into the place of science in human thought. This same trait was characteristic of his text books.

No student of "Old Slick" ever forgot the man who taught him. The anecdotes about him are legion, the apocryphal as true as the actual. Professor Slichter was not always on time and many stories deal with the attempts of the students to capitalize upon this fact and with their subsequent discomfiture. There was the time when a small class on a cold day took to a balcony to watch him enter an empty room. They merely found themselves locked out.

Years later men have found that the shared experience of having been in Slichter's class was in itself a strong fraternal bond. This was the natural outcome of his attitude towards his students. In speaking of his elementary classes he said, "But actually I did not teach freshmen. I taught attorneys, bankers, big business men, physicians, surgeons, judges, congressmen, governors, writers, editors, poets, inventors, great engineers, corporation presidents, railroad presidents, scientists, professors, deans, regents and university presidents. For that is what those freshmen are now, and of course they were the same persons then."

Slichter's scientific work was in the field of applied mathematics and Engineering. His pioneering studies of the flow of underground water has been recognized as fundamental. Largely through his influence the relation between pure and applied mathematics and the integration of the two were on a sound basis at Wisconsin decades before that was the case at any other major American Institution.

As Dean of the Graduate School, Slichter's constructive imagination and strong personality served the University at a time when the problems of growth had to be met by administrative insight of the highest order. This period was one of magnificient achievement for the Graduate School and the many who contributed to this success were always happy to acknowledge the central role of his invigorating influence. The infusion of all graduate work by the spirit of research and development of cooperation in research between individuals, between departments and between colleges received from him a much needed impetus. His force of character and his perception of the means to stimulate scientific work led him to organize the Wisconsin Alumni Research Foundation and his sustained interest was a major element in its success. As a Dean dealing with students, his methods were unusual, even unpredictable, —and perhaps fortunately inimitable. In his hands, they were successful and inspiring.

His interest in Science accompanied an equally deep interest in Literature. He saw in each a rich manifestation of the human spirit. A voluminous reader he was also a writer of rare skill. His essays and speeches some of which are collected in "Science in a Tavern" were always a contribution to the abundant life.

He was primarily interested in the individual. This made him a great teacher and unforgettable friend. It also made him a humanist, a comrade of the men of distinction and of robust character in all ages. He was widely known and had a host of friends but his personality found its deepest expression and greatest source of strength in his family. In the preface to "Science in a Tavern" he wrote: "I dedicate this little volume to the four sons of Mary Louise Slichter; it is these five who, with my former students, have educated me from a crude beginning to a most happy maturity."

Slichter was born in the middle west, lived in the middle west and died in the middle west. He loved this middle west. Its progressive and expanding forces were embodied in his personality. He served its people especially its youth with zest and devotion. The very land, its streams, its lakes, its pleasant countryside brought him contentment. But he was also a citizen of all ages and climes and the inheritor of the intellectual life of both the ancient and modern world. None of his services to this State was greater than the forceful realization he brought to many that the heritage of the past is ours. He once said, "Remember you are all mentioned in Homer's will."

Slichter was a reverent iconoclast. As an iconoclast he liked to startle and to shock. He enjoyed stirring up those whose virtues he admired but whose self-satisfaction he believed was adequate. The episode on the Cambridge bus was typical. As it approached Harvard Square, he asked the driver in the voice we all know carried well, "What are all those buildings"? "Why that is Harvard University." Then as he left, crumpling his white hair, "Strange! Strange! I have lived here thirty years and never heard of it."

But he was reverent in the deepest sense; reverent before the unities of nature, reverent before all great expressions of the human spirit and of the human intellect and especially reverent before the possibilities of youth.

MEMORIAL COMMITTEE
Philo Buck
Edwin B. Fred
H. W. March
Harry Steenbock
H. R. Trumbower
Mark H. Ingraham, Chairman

Chapter 11 ✍ *Family Life*

Mary Louise Byrne

THIS biography naturally focuses on its subject. In the family he had to share the role of hero with four sons. There was however no doubt who the heroine was—his wife, Mary Louise Byrne Slichter.

She was the youngest of nine children of John A. and Maria McKinnon Byrne. The Byrnes were an Irish Catholic family. One of Mrs. Slichter's uncles was a priest in Lismore, County Waterford. On the pulpit of the parish church is his bust, with the following inscription copied by Slichter in 1939: "A monument of sincere regard and undying affection from a devoted and sorrowing flock, to their beloved, genial, and genuine pastor, the Right Reverend Mgr. Byrne, P.P.V.G. who for 32 years zealously ruled over us." The McKinnons were Presbyterians of Scotch descent and some of the children of John Byrne were brought up as Protestants. The Byrne family seems not to have made a fuss about this, but Mrs. Slichter said that her father was regarded as somewhat of a backslider in his family because he married a Protestant. The McKinnons also disapproved of the marriage.

The oldest letter found in the Slichter files in the University of Wisconsin Archives was written in Dublin, Ireland, on May 22, 1802, by Henry Christopher to Ellen Sheehan who later became Ellen Sheehan Byrne, and the mother of Mrs. Slichter's father, John Byrne. There was a note added by Thomas Christopher. We do not know who Henry Christopher was. A suitor? Hardly likely that he would allow anyone else to add a postscript. An uncle? This would possibly explain a reference to his mother. The document must have been precious to Ellen to be kept and passed on to her granddaughters. But what girl wouldn't keep a letter written in beautiful script which said:

~ . . . *We may be allowed to hope that we once again would enjoy the delicious felicity which your amiable Society affords—For me to join in* the common phrase of "nothing new here since your departure" would be departing from that charming principle of truth for which the being blessed with your acquaintance has given me the most pleasing Zest—yes the walks [?] on which I had the pleasure of accompanying you appear to me quite new—the room where I so often met you, the chair on which you rested, the corner of the room where you so patiently & frequently listened to my tiresome and at times Drunken Chat—in a word every thing appears quite altered to my eyes, even the Miss Hart, Miss Brett, Miss Fish, & all the other Misses appear to me as ugly as the D——. The little Birds appear to feel your absence pretty poignantly, they remind me if any thing was necessary to do so, of my melancholy by warbling their little notes in that strain. . . .

Or from the other Christopher:

~ My Ever Darling Girls
    I take this opportunity of Informing you that I feel like a Pelican in the wilderness, since your departure—your departure has given me a Violent tooth ache for I have been unwell since your absence, I believe Grief has been the occasion. There is no remedy I know of can cure me but the happiness of embracing you soon. The happiness I felt while I had you shews me I can't live without your Sex's amiable Society. Therefore I am inclined to enter into the Holy order of Joseph after I consult you which will be 'err long.

The first three children in the Byrne family were born in Ireland; the remaining six in Wisconsin. The family lived on a farm near Otsego, and later, on one near Blooming Grove, where Mary was a little girl. She was born, I believe, on October 15, 1863, although there is some evidence that would add to or subtract a year from that date. She was somewhat secretive about her age—her birthday parties were referred to as celebrations of Virgil's birthday.
    She later described with delight her early childhood on the farm. Her son Sumner has recorded this:

~ . . . Mother was very fond of the horses. She was forbidden to go in the barn but she frequently disobeyed. A colt of which she was very fond kicked her quite hard on the chest. Mother never said anything about it to anyone. Uncle John saw it happen.
    Mother got some yarn and went into the pasture one day and got an old horse. She led him to the fence. She wrapped the yarn around his neck and got up on the fence and got on the horse. Then she rode him around the field. The yarn was supposed to be reins and bridle.
    Aunt Ag and mother played much together but when they did some-

thing wrong and got into trouble, mother had the responsibility of tell-
ing.

Clearly, her quiet but firm independence developed early.

When she was about six the family moved to 446 West Wilson Street, Madison, where the rest of her girlhood was spent. This house remained the headquarters of the Byrne family for years after her marriage.

There being no Presbyterian Church in Madison, the family joined the Grace Episcopal Church, which was a congenial religious home to Mary Louise throughout the remainder of her life.

We have practically no record of her young girlhood in Madison. She entered the University of Wisconsin as a freshman in 1880, having already taught country school near Oregon, Wisconsin. She however left at the end of the year, as did approximately half of the about twenty girls in this freshman class. Except for some special work taken a little later, this was her only year in college. For a year or two she taught at the Fourth Ward School in Madison and then was made principal of the First Ward School, later renamed the Washington School. She held this position until her marriage. (The record of the School Board, which is as above, differs somewhat from her son Sumner's account of a conversation with her in 1946. For three years she appeared in the school records as "Lulu" Byrne, but after she had been principal for awhile, the record was changed to M. L. Byrne.)

Sumner's memorandum contains the following description of this teaching experience:

~ *Mother taught at the Eighth Ward School for six or seven years—*
right through the Friday before she was married. She was married on a
Tuesday. After one year of teaching sixth and seventh grades, she was
made principal. She does not think that she should have been allowed
to teach because she says that she knew nothing about it. She denies
that the family had any political pull but she admits that she knew every
member of the school board. Judge Carpenter, the chairman of the
school board, was the "affectionate" type—the kind who would put his
arm across your shoulders. He used to visit the classes. He visited the
school after mother became principal and asked her how old she was.
She said that she was just twenty or not yet twenty, I've forgotten
which.

Mother says that she is surprised that she was allowed to hold her
job as principal, because she led a gay life, sometimes going to dances
every night of the week. There were dances at the Odd Fellows Hall, the
Park Hotel and a Music Hall at the University. Each fraternity was al-

lowed so many dances a year. After she became engaged to father she never danced. He did not know how to dance and did not care for dancing.

Although she always belittled her pedagogical ability, the evidence both of her personality and the testimony of Max Mason, who was one of her students, make it clear that she was an excellent teacher.

We do not know when or how Charles Slichter met Mary Byrne, I have my guess.

He was baptized an Episcopalian and probably grew up in that faith. Habit, loneliness, a friend, or—less probably—piety, may have enticed him into Grace Church. The annals of the Church are more complete concerning baptisms, marriages, and funerals than memberships; hence, there is no record of when he joined this church to which he belonged for the rest of his life. In a registry compiled about 1871, the members of the Byrne family are listed under an entry of that date. Mary Louise is given as seven years old. The Slichter marriage, in the home of the bride's parents, on December 23, 1890, is recorded; and the rector of Grace Church, Fayette Durlin, officiated.

Although the date when Slichter joined Grace Church is not known, I would guess that he met his future wife through the church where she was a very active member; but of course they may have met earlier, and he may have been persuaded by her to join the church. There was probably a causal relationship but one cannot determine in which order.

This is a case where the course of true love seems to have run smoothly for about sixty years—from not later than 1888 but not before 1886 until Slichter's death in 1946. Those who had seen him only in his moments of anger or irritation thought that his wife must have been a remarkably patient woman. Doubtless she was, as any woman must be whose family consisted of an energetic husband and four vigorous boys; but it is also true that he was a devoted husband.

There are still extant a few notes written by her to him prior to their marriage. He saved letters more than she did.

The following letter shows that she knew him pretty well by the summer of 1888. The formal salutation and closing would have seemed natural to people at that period and do not preclude the possibility that they were already close friends.

Madison, July 19, '88

~ Dear Mr. Slichter:
    I shall be very much pleased to drive with you around Fourth Lake. I think I shall be able between now and then to "hook" enough for lunch.

*Only I know that the prayers of the wicked availeth not, I'd sug-
gest to you to pray for a clear evening.*

I'll have to do it all myself.

Sincerely,
Lulu Byrne

By March 1889 he was taking her to meet friends and was requested
to bring his "fiddle," which his son Louis assures me was a banjo.

I would judge that they became engaged in April 1889 and soon
thereafter he had the mumps, which resulted in his receiving a series of
love letters. By now he was "Charlie" as we see by the letter of May
first:

~ *My dear Charlie:*

*Did you want me to call you "Pat"? If 'twere someone I didn't hold
dearer than anyone else on earth—perhaps—but you have stolen my
heart too completely for anything so frivolous.*

*I haven't gotten over being astonished yet. It never dawned upon
me that I stood at the head—indeed I always felt that if I wasn't at the
foot, I was not far from it. . . . But now I feel that I could fly.
I haven't a doubt concerning my own happiness, and if I felt as well assured of yours, this would be a veritable heaven. . . .*

*My heart went clear down—I don't know how many degrees below
the freezing-point—this morning, when Jennie D. told me you were
not as well last night, but at noon, Mrs. H. told me you were
feeling better. I don't know what the children think of me—half the
time I haven't the remotest idea what I'm doing.*

I have a pile of papers to look over, so shall say farewell.

*You won't venture out, will you?—until the doctor assures you 'tis
perfectly safe.*

*I think you are awfully mean. How can I send you any love, when
you have taken it all?*

Lulu B.

A postscript informs us that she had just spilled a bottle of ink.

By her handling of the "Pat" matter, she showed skill in getting her
own way while building another's ego. To some degree he too got his
own way, for as mentioned in Chapter 2, the version of "Two Hundred
Years of Pioneering" written for his grandchildren was signed "Pat
O'fessor."

There was a letter on May second. The letter on May third should
be of interest to the medical historian, for she wrote: "Now I'll tell you
what I'll do—if you'll agree to drink all the cream and beer the doctor

prescribes, and not let anything or anybody worry you, I'll give you a ride just as soon as the doctor says you can breathe the fresh air."

Christmas vacation took him away again, probably to see his family in Chicago, so there were additional letters, one in particular giving advice concerning New Year's resolutions: "I know you've made up your mind not to bother anyone, haven't you? Anyway I think you'd better."

Spring vacation caused more writing which showed increased knowledge of what "My dear Charlie" was like: "I wonder what you have been doing all day. I don't believe you went to church. You're probably succeeding in wearing everyone out. Will you admit it?"

They were married December 23, 1890. The groom's mother could not come to the wedding because of poor health, but welcomed the bride in Chicago shortly thereafter.

1890–1909

From their marriage until he started surveying western water resources, about 1902, he was seldom away from home for a long period. Hence, there are few family letters in those years.

Upon their marriage they moved to 636 North Frances Street in Madison, and this remained the family home until Mrs. Slichter's death in 1955.

Fifty-one years later, on December 21, 1941, Slichter wrote his son Louis:

~ This year is only 51, and we are now beginning counting backward 51 years at 636! None of my colleagues on the faculty have such a record and none have a record of such wonderful boys. That long residence at 636 has hallowed it to us and all our memories are associated with the simple rooms and meagre adornments of that Home— for such it has been and still is. Your mother stepped into that home as a bride and all the children were born in that front bedroom. No wonder it is not for sale

The first decade saw the births of their four sons: Sumner Huber, January 8, 1892; Louis Byrne, May 19, 1896; Allen McKinnon, February 18, 1898; and Donald Charles, July 3, 1900. (I will, in general, refer to them by their first names.)

Perhaps there is no better place to record the accomplishments of these boys than here.

Sumner became one of the country's most eminent economists,

holding a Lamont University Professorship at Harvard from 1940 until his death on September 27, 1959.

Louis took his Ph.D. in Physics and became a geophysicist. For years he taught at the Massachusetts Institute of Technology, then at the University of Wisconsin, and finally at the University of California, Los Angeles, where he headed the Geophysics Institute until 1963 when he technically retired—although he continued productive research. The Institute building at U.C.L.A. is named in his honor.

Allen is a successful businessman with the Pelton Steel Casting Company of Milwaukee, becoming its president in 1947 and Chairman of the Board in 1962. He has been an active, intelligent, and generous citizen of the city and the state.

Donald also went into business—most of his career being with the Northwestern Mutual Life Insurance Company, becoming president in 1951 and Chairman of the Board in 1965. He has been amazingly effective in the educational affairs of the state, including the University; and his wise counsel is frequently put to the service of his fellowmen.

All four sons graduated from the University of Wisconsin and married graduates of the same university.

These few sentences do not portray the true distinction of the brothers, a distinction which, incidentally, is reappearing in the next generation.

The boys grew up in an atmosphere of trust and affection. There is both internal evidence and their testimony, as well as that of visitors in their home.

Mrs. Slichter's nieces, when they went to college, frequently visited their aunt and uncle, enticed by good company, good food, and an occasional buggy ride into the country. One of these, Agnes Merrill (now Mrs. Holton H. Scott), who attended the University from 1898 to 1902, was a great favorite of the family. She recently wrote to Donald giving her memory of his father. Her first paragraph reads:

~ *As I think of Uncle Charlie, the name we relatives always used, the characteristics that stand forth in my mind were his interest in everything worth while, especially to University affairs, his kindness to those of us he liked, and above all his devotion to his family and wife.*

She further relates:

~ *In their family life he left most of the discipline and training of the boys to Aunt Lou unless it was a matter of character. I recall many times at the dinner table when the boys would become too boisterous, and Aunt Lou would ask his help. Then he would say, "Now see here,*

Fred, enough of that. Your mother wants order." Fred was a sort of generic term that the boys understood meant all of them, thus saving individual names.

We have a somewhat different account of the family discipline from Louis Slichter who on April 25, 1970, in a conversation with me, stated about as follows:

~ Of course, father thrashed us sometimes. His weapon was either a slipper or his razor strop. Most often it was because we had gone on the ice when it wasn't safe. He seemed to want to preserve us. Moreover, when any one of us was punished, the other three enjoyed it so much that I guess the total effect was to give more pleasure than pain.

One time when the dean was showing his well-equipped workshop to me and my seven-year-old son, whom for the day he dubbed "Patriarch," he said to us: "I didn't want the boys to be extravagant but I did want them to know how to use tools. I told Wolff-Kubly [Wolff, Kubly & Hirsig Hardware Co.] to honor any order for tools that they gave."

Since he kept the tools his generosity may be explained in ways other than the one given. Moreover, Louis says that he and Don were the only ones who cared to work with their hands, and that neither of them would have dared to abuse the privilege of ordering tools. Incidentally, he kept his tools, nails, and screws in a much more orderly fashion than his letter files.

The legends of Frances Street would indicate that producing four distinguished citizens was not accomplished by making them model boys. In her autobiography, Gertrude Slaughter, in speaking of certain joyous occasions of her old age, says, "Another was the convocation called to honor the 'Slichter boys,' who had once been the clever little bad boys of our neighborhood." In fact, I am told by Edwin B. Fred, there was a period when the Slaughter daughters were not allowed to play with the Slichter sons, lest their vocabulary be unduly enriched.

When the man of the house began to travel, we begin to get more contemporary accounts of the scene at "636," at least as it was when he was away. There were some days when life was not a "veritable heaven," for on July 7, 1902, Mrs. Slichter wrote her husband: "This is a terrible morning. We have had three days of constant rain, so the children have not been able to go out at all. All they do is to quarrel all day long. My head is sore."

About this time she also reported: "Louis and Allen are good as usual. Donald makes a great row at not being included in everything, and rises in his wrath when I leave them. Sumner is no help to the situation. He keeps one under a severe tension the whole time, but these I suppose are a parent's reward."

Perhaps one of the occasions of Donald's wrath was when the father took Louis and Allen to Chicago. In a memorandum of August 30, 1946, Slichter gives the following account of one episode in this trip:

~ *This history takes us back about forty-four years. At that time I had taken my little son age 4 and his brother aged 6 to Chicago to see the Field Museum, then in Jackson Park; to visit the toy department at Marshall Fields; and, as a climax, to see and hear the President of the United States, then the hero of every child. After spending several days inspecting the ancient railway trains, the stuffed animals and the "stuffed Indians," skilfully arranged in tribal, and family groups, we prepared finally for the great day. Our hotel was at Jackson Park, so it was an easy walk to the University of Chicago campus, where President Theodore Roosevelt was to lay the cornerstone of the Law Building, receive an honorary degree, and make a speech. With the skill I had acquired as a college student, I took my small son on my shoulder and successfully wormed my way to the very first row of the crowd. The cornerstone was laid, the degree conferred, and the President stepped forward to address the crowd, splendidly adorned in his purple trimmed gown and hood. The crowd came to silence, only to be shocked by a little voice from my shoulder which suddenly piped up:*

"Is that Prezdent Rosefelt in the purple sweater?" My face turned red with embarrassment, for I knew President Harper well and I should have been known to President Roosevelt. But my embarrassment was soon over, for both President Harper and President Roosevelt broke out in such robust laughter that even the solemn row of trustees seated on the platform began to smile. That innocent appraisal of academic costume put Roosevelt in accord with his audience—and is worthy of a place in history.

When he sent Allen his copy of this footnote to history he added, "This was a glorious trip but you have probably forgotten all about it. This is the official record for your memory file. Faithfully and affectionately, Fessor."

It was not many years after Mrs. Slichter's plaintive letter, before Sumner became more of a help. On March 22, 1905, he wrote:

~ *Dear Papa:— All is well. We are very careful about the Lake. Monday Louis went to Shurly's for supper, and yesterday Allen went to La Follette's. He didn't stay for supper, and it wasn't a birthday. Seven other children went, among them Malcolm. They had lemonade and* cake. Allen enjoyed it very much. The picture is for mama. Allen made it. It is a picture of a battle between the French and the English. The men on the left are the French. I got 100 on an arithmetic test. I am forced to announce that you owe me ten cents as I saw a robin in Munro's apple tree this morning. I have heard a crow. Allen always sets the table. He seems to enjoy it.

With love
Sumner

Sumner's scholastic prowess in mathematics was almost matched in spelling for he reported to his father—"I got 96 in a spelling test of fifity words."

Also on March 22 Donald made his report, Sumner acting as scribe.

~ Dear Mama:– Deary Papa:– I'm not going on the lake. I'm not going in the snow. Sumner write the letter. Aunty Ag is very good to us. I was up in Sumner's room watching him write the letter. I told him what to write. I am going out of doors on warm days. Hugs and kisses. I didn't go up in papa's study. I'm a good boy. I got mama's letter.

The lake seems to have been a constant worry especially to the mother.

It was about this time that Sumner wrote his father, "The boys have been good. Louis the least." Perhaps one can understand the "Louis the least" from Louis's letter to his father written September 4, 1906. (The original spelling is retained.)

~ Dear Papa:–

Malclom, Allen, Paton, and myself rote some signs that said there was going to be a ~~eigr~~ circus at the Hofeltʃ boys house and are going to pin the signs up on the St. and a lot of people will come and there ~~wau~~ will not be a circus. Each sign read:—

NOTICE
A circus will be Held at 4 sharp Wednesday afternoon at 621 Frances St. For Boy and girls and Grow Ups.

I shall write mama next. I have just come home from Malcolm's. I was looking at his time tables. I must stop here because must go to bed. Shall Donald go

to to school.   ~~Ee~~ Elizebeth and Juili ~~Julia~~ are   With much love
goieing.   Louis

"Malclom" was Malcolm Sharp, son of Professor Frank C. Sharp; Paton may have been Paton McGilvary, son of Professor E. B. Mc-Gilvary; "Hofelt boys" referred to the sons of Professor A. R. Hohlfeld.

By March 25, 1908, Louis's spelling had vastly improved, but his sense of humor was still intact. While visiting an uncle in Kansas City he wrote to his father:

~ *Here is a joke Lucia told—there was a little boy about 4 years old and his mother didn't like to whip him. One day she made up her mind she would whip him. Just as she was raising her slipper he said, "Go nice and easy, little mother." He didn't get the licking.*

The boys' father took much interest in their activities. For instance, he saw to the typing of Allen's arrangement of Homer's account of the fight between Ajax and Hector, for acting by his three youngest sons. Louis was Hector; Allen, Ajax; and Donald was the Herald of the Gods.

We find him sending Louis to take a manual training course in the summer of 1905. It was probably somewhat after this that Louis reported his mother had been making fudge, and that he had been making a new boat. Both occupations seemed to him to be appropriate.

For a period the boys had a passion for collecting, especially for collecting stamps and timetables. For instance, Allen, in August 1904, dictated a letter to his father:

~ *Dear Papa:– We are having a good time. We are being good boys. Please send me and Louis a Mexican Central R.R. time table. Are you having a good time? Is it pretty were you are? Louis bought a fine sail boat for 25¢. He found the money in our yard. The boys are going to have toy boat race.*

Louis and Malcohm have a fine museum. They have some star fish and some glass from the Chicago fire. They have some wood and bark off of the oldest tree in world. Malcohms Aunt Annie gave them a lot of fine things. They have a live crab. We have been out swiming and we had lots of fun.

Love and Kisses from
Allen

(The quarter was not the last thing found on the Slichter grounds. Years later Professor E. N. Hiebert, who had lived in the Slichter Frances Street home after it had been given to the University by the Slichter sons, was able at a dinner to present ceremoniously to Charles Pence Slichter the Phi Beta Kappa key of his grandfather, Charles S.

Slichter, Northwestern, 1885, found in the rose garden of that same backyard.)

On June 8, 1909, Slichter wrote to J. W. Winterbotham, Secretary of the Railroad Commission of Wisconsin:

~ I have your envelope, enclosing a large number of stamps for my boys. I write to thank you for these. I wish you could see what great pleasure they have given the little fellows. Your thoughtfulness has given them almost as much pleasure as if you had sent them an automobile, as they are intensely interested in stamps and in the promotion of their collection in their coming year abroad.

But the day when the boys would wish a car would come. Slichter wisely preserved in his papers the following undated agreement:

~ We, the undersigned, hereby agree to keep our auto shined up and cleaned, each according to our share.

Signed

Louis B. Slichter

Donald C. Slichter

The Right Honorable Allen Slichter, Esq.

The Year Abroad

If Slichter's publication of "Theoretical Investigation of the Motion of Ground Waters" in 1899 marked the peak of his scientific career, and the launching of the Wisconsin Alumni Research Foundation in the period around 1925 his greatest administrative accomplishment, no year would be richer in the memory of the family than 1909–10 when all of them, at times accompanied by friends and relatives, were together in Europe.

On June 11, 1909, a family milestone was reached when Sumner graduated from the Madison High School with a fine record. Another member of his class was Nellie Ada Pence who, nine years later, was to become his wife.

The younger boys were also old enough to get much from a European trip.

At that time there was an arrangement at the University of Wisconsin by which one could teach in the summer school, waive the immediate receipt of salary, and, later during an academic year, receive pay while on leave for a somewhat longer period than the summer school. One would also get the benefit of salary increases which had occurred in the meantime. Slichter was entitled, on this basis, to three-

fourths of an academic year on full pay, and in light of his uninter-
rupted services the Regents, on recommendation of President Van
Hise, gave him an additional half-semester's leave with pay. This was
poetic justice in advance, for later, as graduate dean, he would secure
similar opportunities for many faculty members.

The basic plan for the year was to have a short visit in England, then
to proceed via the Rhineland to Switzerland for a hiking trip, to be
followed by a winter in Munich where he and the boys would study.

Except for Sumner, the family, as well as Mrs. Slichter's niece,
Agnes Merrill, who at the time was a teacher of Latin, sailed from
Montreal for Glasgow on the Donaldson line on July 29, 1909. Sumner
followed somewhat later, sailing directly from Philadelphia to Holland.

On July 19 Louis started to keep a diary in a small pocket notebook
which allowed seven two-inch lines to each day. The entry of Saturday,
August 21, is typical, "Went to Delft and saw a lot of china. Rainy.
Saw leaning tower and the Prinzen Hoff. Saw a lot of canals. Left The
Haag at 4:18 and got to Amsterdam at 5:30. It is full of canals."

The early portion of Allen's diary was less succinct. The entry of
August 21, the same date as the quotation above from Louis's diary,
was [spelling retained]:

~   It was raining in the morning. Got a train to Delf. Mama bought
some china, made by the famous pottery company at Delf. Saw a
church that was build in the thirteenth century. It was called the "New
Church". Saw lots of canals in Delf. Saw a church that had a tower
that was leaning. It was an old church but not as old as the "New
Church." It was called the "Old Church." We saw Prinzen Hoff, a
small palace where William the Orange lived and he was murdered
there. We did not go inside the palace. We took a steam train to the
Haag. After lunch we went down to the station and got our train to
Amsterdam. We got to Amsterdam at about 5:00 P.M. Papa went to
find a hotel while we got the mail at Cooks. After lunch we bought
lunch for tomorrow, then we took a car to our hotel. It was a private
hotel.

In addition to keeping a diary, Allen made a catalog of some of the
art treasures of London and Munich—an astonishing enterprise for a
boy under twelve years of age. A forecast of his conscientiousness in
money matters, as well as a hint of how the boy's taste was developing
into that of an adult, are given by his expense accounts on the last
page of the first volume of his diary.

~ Expenses

Shoeing the Bay Mare by Lanseer	$.60
A little led dog	$.06
Four Tunis stamps	$.03-¾
A set of stamps	$.07-½
A carved bear	$.35
A German print	$1.12-½
Airship	$.56-¼
	$2.81
I had	$2.59
I owe Papa	$.22

The letter of Mrs. Scott (Agnes Merrill), from which we have already quoted, has the following account of fall and spring periods when she was with the Slichters:

~ The trip to Europe in 1909–10 was a great event in the lives of all of us. As I had planned to visit Rome with Lily Taylor that year when Prof. Slaughter was the head of the Classical School, the Slichters asked me to sail with them and join them in walking trips in the fall and spring. As I read my diary for that period, I realize again what a privilege it was for me to take the trip with Aunt Lou and Uncle Charles and the sons. Uncle C. had a great sense of humor and an interest in everything or everybody we encountered. Both he and Aunt Lou enjoyed the cathedrals, museums, the scenery on our way through Scotland and England. If Aunt Lou and I ever decided to skip something and rest, he would come back and say, "Oh girls you have missed the best things I have seen yet." In Switzerland though we seldom missed anything, it was always the "most gorgeous view."

Uncle Chas. had made a great study in preparation for this trip— the first for all of us—and was always full of enthusiasm. We traveled comfortably but with as much economy as possible. We soon learned to buy our own food and eat it in the parks or in our lodgings. In Switzerland it was always beer, rolls and cheese for the adults with chocolate and milk for the boys. My diary tells of our first impressions of Heidelberg as we sat in a public square to eat the lunch bought at a bakery, a little apprehensive that a policeman might order us to move on.

Uncle Charles made all hotel arrangements. It was always a matter of rejoicing to the boys, if the plan included rooms and meals as then there would be no restraint as to food.

On our walking trips we must have presented as interesting a sight to the natives as they were to us. Heavy luggage was checked. We all carried overnight essentials in knapsacks and we walked never together. Uncle Chas. was always ahead and Aunt Lou in the rear with Louis, Allen, Don and myself in that order between. (Sumner did not join the family until later.)

The Professor liked to do all of the planning, which sometimes irked Aunt Lou. I recall one incident when we were in Switzerland. She said to me "I'd like to see if you and I could plan a trip ourselves. I'd like to go to Paris but Charlie is not interested. Let us go by ourselves and leave the boys with their father." So we quietly inquired about trains and hotels. Then Aunt Lou broached the subject to her spouse who replied, "If you go to Paris, I will certainly go with you." This of course upset the whole project, as seeing Paris with three small boys in tow was not our idea.

He was always fond of a joke. One time in England he had visited an evening session of the House of Parliament. In describing it to us the next day, he showed us where the Prime Minister Asquith had sat; also the Cabinet and the Opposition headed by Balfour, then added Winston Churchill was the "dressy bag."

The boys like to tell of a visit with their father to Garmisch for winter sports. They happened to sit at a table in an Inn where a bürger was drinking beer from an enormous stein. Uncle C. managed to fill a similar stein with water and exchanged the two. The poor man was nearly drowned by his next drink.

As Uncle Charles was very fond of good music, the winter in Munich gave him and the family opportunity to hear much opera and many good concerts. They made many friends through the University. As I joined them in the spring, my diary tells of the "gemütliche" teas or evenings for which they made an appreciative audience though not performers themselves.

Later in Madison the sons presented him with a large victrola which gave such loud music that though he was entranced, chance visitors retired with Aunt Lou to the upstairs study.

Slichter kept a collection of the programs of plays, operas, and concerts he attended. He had a busy year.

Years later Professor Herman March wrote an undated letter to Louis Slichter recalling the days in Munich, where he had also been in 1909–10. One of the episodes concerning Slichter he recounts as follows:

~ He met a pair of German students, one of whom wore the gaily colored cap and diagonal stripe of his corps. This one had a red cap very similar to those worn by Dienstmänner, who correspond to our Red Caps. But the Dienstmänner have numbers on their caps. Your father went up to this student and asked him what his number was, pretending that he had some baggage. The student commenced to blush and his friend to laugh. Your father pretending that he thought that the man did not understand pulled out his guide book and showed him what he meant. Your father said that he never saw a man so red in his life and the friend just shouted with laughter.

This happened in the fall of 1909.

In Munich he found the scientific atmosphere stimulating. The Ludwig-Maximilian University, commonly called "The University of Munich," gave him a formidable-looking document which evidently was the academic equivalent of the "Keys to the City." On December 30, 1909, he wrote a letter to President Van Hise expressing gratitude for the faculty salary increases, including his own, and gave descriptions both of the abundance of Christmas trees in Munich and the Bavarian state ownership of waterpower and forests, which he had come to approve: "If we give away our waterpower we are the darnest fools on earth." He also stated:

~ I am having a very happy year in Munchen. I have become acquainted with some of the very best of the scientific men here and am profiting as much as I can from the opportunity. The work I have done on Groundwaters and my connection with the Reclamation Service has been of the very greatest benefit to me, for I was amazed to find that I was known and my work summarized in their textbooks. In this way I have the opportunity to meet the scientific men and the engineers. This class of men is much more interesting than the German mathematicians, and so my circle is a much larger one than it otherwise would have been.

Even if, as Mrs. Scott reports, a great deal of preparation had gone into the trip, not everything went according to plans. We have a letter to the Collector of Customs of the Port of Boston, protesting the payment of $27.50. After a year away, he was still in good form. He had copies of the regulations which he kept permanently, and he said:

~ It is clearly evident that I have conformed in every particular with the written instructions furnished me by the treasury department and the U.S. Consul. It seems clear that I have a right to expect these instructions, in their fair and reasonable interpretation, to be as binding upon the treasury department as upon me, and not to be in the *position of finding myself with one sort of instructions for the passenger and another for the custom officer.*

Mrs. Slichter did not try to blame anyone else for the fact that she could not use the supply of United States government postcards which she had taken with her to mail home.

Of course no major experience of a member of Town and Gown was complete until the definitive account was recorded in the annals of the Club. On November 10, 1910, the group met at the home of J. W. Cunliffe, Professor of English at the University of Wisconsin, who, after describing in the minutes the desultory conversation of the evening, concludes:

~ *Slichter was then allowed to begin his account of a year in Munich, and with the aid of a running fire of questions delivered his soul of a very lively presentation of life as it is lived in a modern German community, where he wasn't allowed to hang his trousers out of the window, and could hear 80 different operas without bankruptcy. The benefits of municipal government by experts were set forth in such a way as to arouse admiration and envy, and the various features of the educational system were discussed. The opportunities afforded for mathematical research made Slichter's mouth water, even in retrospect, and by the eye of the imagination he was seen enthroned in a magnificent palace on Lake Mendota, surrounded by mathematical minions innumerable, and encompassed with gorgeous apartments specially designed for the devotees of the queen of the sciences. A second palace, showing what mathematics had been from the days of the mound builders 'till now, was more dimly foreshadowed. It was remarked that Slichter said nothing about beer, but a rigorous cross-examination failed to elicit more than the information that Munich was a place where beer was to be had.*

Long Lasting Family Interests

Some aspects of the family life are of such a nature that they should receive separate descriptions but lasted so long that they do not fit in any of the chronological divisions of this chapter; hence we interrupt the sequence of the story, somewhere near its midpoint, to consider them.

Relatives. We have already quoted from Slichter's correspondence with his brother Fred concerning the theory of evolution, and later we refer to the purchase of a baby carriage for Sumner through his brother-in-law George. His brother David, who lived in South Dakota, named him his executor—a job that turned out to be far from light. The small bequest he got from the estate was offered to the University of South Dakota as a loan fund for worthy students. His uncle John [?] consulted him in 1893 concerning the originality of a trivial arithmetic method which the uncle took seriously. It is clear that the youngest was respected by his elders; also by his juniors, since nieces sought from him financial and educational advice.

In 1943 Slichter gave pictures of his own mother and father to his son Sumner. He described these pictures as follows:

~ My recollection is quite clear that these pictures were taken on glass plates by Hassler (the photographer of Lincoln) at Galena and the prints sent to John Sartain at Philadelphia for coloring and painting. I have also heard that Hassler (or Hessler) was one of the first in this country to pass on from Daguerreotypes to glass negatives which was a process he learned on a trip to Europe. Glass plates were first used in 1851 so the date of the portraits lies somewhere between the time my father left Galena and 1851. I do not know the date that my father left Galena. He was there in 1854 when the church record shows that my brothers were baptized. I shall try to locate the date more exactly.

Hence the evidence we have indicates that there was mutual affection between Slichter and his parents and brothers and sisters, but that the family was not an extremely close-knit clan as were the Byrnes, or the family that he and his wife reared.

Slichter did not see the members of his own family with anything like the frequency with which he and his family saw the Byrnes. The Byrnes' headquarters was Madison. Mrs. Slichter's mother died in 1894, when Sumner was a baby, so that none of the boys remembered her; but her father, John A. Byrne, lived to be eighty-nine, dying on March 1, 1910, while the Slichters were in Europe. His grandchildren knew him well and had great affection for him.

Two of Mrs. Slichter's sisters, Agnes and Isabel, never married, and continued to live in Madison until their deaths—Agnes at the age of about sixty on June 6, 1922, and Isabel on October 3, 1935, at the age of ninety-three.

Perhaps of all the uncles and aunts, Isabel Byrne, "Aunt Belle," made the greatest impression on the nephews. She was born in Ireland in 1842, the oldest of the family. (I assume that Mrs. Slichter's statement to her son Sumner is more accurate than the church record, which would indicate Isabel was somewhat younger and not the eldest child. Mrs. Slichter's memory seems to have been far less confused than the registry of Grace Church.) For a quarter of a century Isabel taught in the public schools of Madison with marked success in controlling her pupils of Irish descent. When the family was in Europe in 1909–10, "Aunt Belle" kept track of the Carroll Street business, showing a firm grasp of its details. After she died, his father wrote to Louis:

~ *I wrote you a brief note telling of the passing of Aunt Belle. Altho her early friends had all preceded her and she was, in a real sense, the last leaf upon the tree, yet it was a satisfaction to see in what esteem she was held by so many of the present generation. Her ancestry on both her father's and her mother's side were cultured and sterling people and Aunt Belle had all the admirable qualities so valued in gentlewomen of three generations ago. From the wonderful companionship of your mother which has been mine for nearly a half century, I am in a position to know of the uprightness and courage and affection that is possible to one of her family. In her last days she spoke in such affection of Louis' "wee girls" as she called them. To her all of her relatives, near or distant, were the salt of the earth. In early days, as I knew her first, she was really the hostess of 446 West Wilson Street and no one knows better than I of the intimate and cultured hospitality she could extend to a diffident stranger. As an insignificant university tutor I was not a little surprised that she found something worthwhile in myself and that she was quite willing to smile on my ardent and glorious courtship of your mother. Now all is quite as real as ever, although it is only seen in reflection in memory's mirror.*

Louis replied, in part, "I'll miss her greatly, for there was no one like her—at least no one left that I knew. She belonged to a fascinating period in the development of Madison and the country, and seemed to remember everything. And I guess there were no finer representatives than she of the pioneers of those days—a cultured lady of the highest type."

It was Uncle John Byrne, I am told, who on his visits always greeted the boys with, "You four young scalliwags need a good sound thrashing." Even when he tried to look stern, his known kindness kept his nephews free of fear.

Two of Mrs. Slichter's sisters—Ellen Byrne Merrill, "Aunt El," and Maria Byrne O'Dell, "Aunt Mide"—were considerably older than Mrs. Slichter. The four Merrill daughters and the two O'Dell daughters all came to the University, being there between the years 1894 and 1908. There was only one year in this period when no one of these nieces was in attendance. They were welcome in the Slichter home and were glad to accept Aunt Lou's hospitality. They must have given her support in diluting the unadulterated masculinity of her fold.

One of these nieces was Agnes Merrill Scott, whose account of at home and abroad with the Slichters we have already quoted. A sister, Grace Merrill Hodgkins, also kept in touch with the Slichters for a long period. From 1939 to 1950 her husband, Walter Hodgkins, served as a regent of the University of Wisconsin. There is some suspicion that Frances Street had been influential in pushing his appointment by Governor Heil. If so, it was not the first time Slichter had taken a hand in the appointment of regents. He seems at least to have worked (unsuccessfully) for the reappointment of Hodgkins's father-in-law, and against (successfully) the appointment of a woman whom he felt was rather long on nuisance value. In 1910 he pushed, through friends, the appointment of Thomas E. Brittingham, I judge with the approval of President Van Hise. There is little doubt that the new Regent discussed the University affairs with the Dean Emeritus.

The only close relative on the Slichter side who lived nearby was his nephew, the well-liked Laurence C. Burke, a member of the staff of the University of Wisconsin Library from 1902 to 1947, ultimately associate librarian. It was he, who with the librarian and the aid of a small horse cart, moved the books of the University of Wisconsin Library from what is now Music Hall to the State Historical Society Building in 1900, and returned in 1953 to take the Coverdale Bible from that building to the new Memorial Library as a token of its opening. In 1953 professional movers transferred the rest of the collection.

Food and Drink. There is no doubt that all the Slichter family liked food and liquid refreshments, although Mrs. Slichter effectively discouraged finickiness. I judge they were hearty eaters, but very moderate drinkers; however, they discussed each with equal gusto if not with equal frequency. For instance, "We were also given a baked boneless ham by Mr. Bolz of the Oscar Mayer Co, so it looks as if we might need another bottle of Scotch to avoid wasting food."

If Scotch whisky was a lubricant for ham, Scotch oatmeal (Scotch Douglas—he also had quotations for Grant's Scotch Steel Cut Oatmeal from Dundee) was a fundamental component of the Slichter diet.

His oldest grandson when a small child used to run over from his father's house to Sladshack after breakfast for supplementary snacks and would find 'Fessor before a huge, well-filled bowl, large enough to contain much oatmeal and about a pint of cream, partaking of the contents with evident relish. When the cream outlasted the oatmeal

the remainder was added to the coffee. One must remember that Mary
Louise Byrne had a McKinnon for a mother.

The delight that the contemplation of a sumptuous dinner gave
Slichter is shown in a number of his essays—more than any other in
"The Royal Philosophers," the first in *Science in a Tavern*. Slichter
was not only delighted with the menus of the Royal Philosophers but
correctly assumed that the members of the Madison Literary Club
would also be, and listed four in detail. Perhaps the most modest was
the substitute dinner when the expected turtle expired in transit:

Turkey Boiled and Oyesters	Two dishes Herring
Calves' Head, Hashed	Tongue and Udder
Fowles and Bacon	Leg of Pork and Pease
Chine of Mutton	Sir Loin of Beef
Apple Pye	Plum Pudding

Butter and Cheese

Those whose imaginations demand something more hearty should
read the whole essay.

His letters to Louis and Sumner tell of the feasts, especially the
Christmas turkey dinners at Donald's and Allen's homes in Milwaukee.
He also tells of finding good places to eat in Middleton, Dodgeville,
Fort Atkinson, and Baraboo. The vegetables from the garden at Sladshack, particularly the fresh asparagus, were a recurrent source of
pleasure. The grandchildren's name for Mrs. Slichter—Grandma Cakes
—is a tribute to her artistry, and places eating in proper perspective.

The members of the intermediate generation were also trenchermen.
Many of the letters from Slichter's sons to their parents show their
appreciation of the table. I have already quoted in Chapter 4 the 1905
letter from Sumner to his father detailing the menu at his grandfather's
birthday party.

Of course during a walking trip, at college age, the appetite is at a
maximum. The following is from a letter from Sumner to his mother
dated August 26 (I think the year was 1912 although no perpetual
calendar will tell you in what year Monday arrives both on August 11,
as in a previous letter, and on August 26.)

~ *This mountain air in connection with hard walking is awfully
hard on your appetite and I am eating worse than ever. We had pretty
good luck lately in getting hotels where you get a lot—especially at
breakfast where it is hardest. When they only bring you one or two
rolls at a time it doesn't do much good to order more. The best meal we
have got so far was at the pension at Grindelwald. It cost 2.50 frc.*

*which is more than we usually pay for dinner. First they brought us a
big bowl of soup which held two soup dishes apiece. With that we had
a hunk of bread about as big as a man's head. We ate it all. Second
course was ham and eggs—two eggs and three slices of ham apiece. Next
was tongue rolled in crumbs. The platter was heaped full of fried potatoes. I must have had four helpings of potatoes. After that we had
lettuce salad and cake with brandy sauce. I don't see how they served
it for 2.50. I'm glad we're leaving Switzerland in one way because we
would soon be broke eating.*

Nearly thirty-two years had not diminished his interest in food nor
his faith in his father's similar interest, and perhaps they had increased
his love for a bargain, when on June 2, 1944, he wrote:

~ *I dropped into the Statler Hotel in Detroit yesterday afternoon to
buy a paper. A big sign in the lobby said that anyone buying a hundred
[dollar] war bond at the assistant manager's desk would be given a
free dinner. That made the bond look like a fairly good security since it
raised the yield from 2.9% to about 3.5. I had a spare portrait of General Grant with me and also a couple of extra portraits of President
Jackson so I bought myself a bond and had a delicious white fish dinner. Next week when I return to Detroit, I'll take a couple of hundred
dollars with me and if the offer is still on, I'll eat free at the Statler. I
must buy several bonds in connection with the drive anyway and I
might as well buy them where they are good for food.*

On May 16, 1943, Donald describes his visit to Louis and Martha in
California. "She had a wonderful dinner, roast beef, etc. which she
had to fix up herself without any help."

On June 6 of an undetermined year Allen wrote "Mom and Dad":

~ *That asparagus they [Donald and Dickie] brought from you last
Sunday was most welcome,—and we enjoyed it immensely. Thanks so
much for such excellent fodder! Did you hear how near we came to not
getting it? Don brought it in from his car and put it in their sink.
Dickie thought it was garbage and deposited it in the garbage can.
Later the error was discovered, but no one was squeamish and it was
not damaged in the least.*

On April 14, 1939, Louis wrote to "Pop and Mom":

~ *We had a very fine, though brief, visit with Al on Thursday—he
arrived Wednesday night and we took him to the Old Oyster House,
(Union Oyster House) for a good lobster dinner. Too bad Sumner
couldn't have been present. Ada, Martha, Al & I were the four. We*

*had oysters, clam chowder, broiled lobster, lobster salad, & beer &
desert @ $1.50, which I believe is a better figure than could be quoted
at Madison.*

It was, if I remember correctly, in the spring of 1935 that Louis took
me for lunch to the same oyster house; and, knowing that I would be
for some time away from the land (or rather the sea) of the oyster,
gave me raw oysters for a start, fried oysters for the main course, and
oyster stew for dessert.

It can perhaps be gathered that my approval of the Slichters' attitude
towards food is unbounded; in fact it has been hinted that the emphasis of this section is as much mine as theirs.

Sladshack. There was one spot that came to mean more to the
family than any other. In 1892 Slichter and his wife, individually,
acquired two lots on the northeast shore of Lake Mendota. To start
with, he moved a boathouse across Lake Mendota on the ice to serve
as a shelter. Successive purchases added to the grounds. He built a
small cottage, which in stages developed into a winterized home with
a workshop and boathouse nearby. He also had a hen house, and a
garden which provided sweet corn and asparagus and other greens. For
awhile there was a pigpen with at least one occupant. One year, in
addition to "Mona Lisa," there was a foundling runt left by an anonymous donor. The pen gave way to a tennis court. Later, Sumner had
his summer home on adjacent land.

There were many native wild flowers on the place and Slichter added
to them. They grew profusely. He was delighted that the blooming of
the bloodroot would celebrate his birthday. According to his account,
its punctuality almost rivaled the swallows of Capistrano. The trillium
parties came in May.

The cottage was called "Sladshack" for Sumner, Louis, Allen, Donald. In its early days there were steamboats which delivered persons,
goods, and mail to various points on the lake. In summer, Slichter's
address was merely "Marine Service, Lake Mendota, Madison, Wisconsin." The boys with their bicycles and their dogs could wander at
will over the countryside. By modern standards the roads were poor and
certainly uncrowded by cars. The telephone service was so meager that
it hardly afforded communication. It certainly did not afford privacy as
it was a ten-party line. (The reaction to this situation was disclosed in
Chapter 7.)

Slichter reveled in the sunsets. "Eagle Heights is a dark blue and the whole glare is on the ominous side of beauty so characteristic of autumn evenings."

As the boys grew, Sladshack became the center of much fishing and sailing, later shared with grandchildren. Sumner's son Bill received a membership in the Mendota Yacht Club from his grandfather.

A place such as this had to be defended in various ways. The three chief hazards appear to have been fire, theft, and pollution. On April 26, 1923, Slichter wrote to Sumner:

~ I just saved the boat house and the shops at the shack from the flames yesterday by the skin of my teeth. Crowley's new nut had started a fire and it had got away from him. I got to the shack at 1:30 and found a regular prairie fire. I grabbed a shovel and told the nut to get a spade—he was fighting with a rake which only made things worse. We stopped the fire before it reached the chicken yard. It had travelled ⅓ of the tennis court and within 50 feet of the boat house. I am afraid that if the boat house had caught fire that it would have spread to the cottage as so many leaves on the bank would have been too much for us. I was completely exhausted. I worked so hard and am not over it yet. Things had gotten dry awfully soon after the last rains and I am very much afraid of fires. It probably ruined some small apple trees on your lot I had been nursing for three years.

More than once thieves got into either the house or the workshop.

The nearby state hospital for the insane was accused of allowing the lake to be polluted. On July 24, 1939, he wrote to the Commissioner of the State Board of Control and to the State Conservation Commission.

~ I have been much interested in the State's efforts to attract visitors to our State and to preserve its natural beauty. One of the beauty spots that many visitors see and one of the most attractive in the State, is, of course, Lake Mendota at Madison. Fortunately considerable stretches of the shore of this lake is owned by the State itself. One of the most beautiful spots is so-called "Governor's Island" on the grounds of the State Hospital at Mendota. Unfortunately the recent Secretary of the Board of Control and the Board itself and the Superintendent of the State Hospital have never inspected this spot. They have permitted (perhaps by neglect and not by approval) a number of ugly shacks and wretched cabins to be erected at various points of the beautiful shore, so that this beauty spot looks as bad as the terrible jungles along a neglected railroad yard. They have permitted trash and all sorts of discards such as bed springs, old stoves, tin cans and auto tires to be dumped along a beautiful piece of marshland where water fowl have for ages made their nests. This shore line was at one time beautified by the building of a roadway around Governor's Island and the planting of many shrubs and groves of white pine and other trees. This work was done from funds raised by the late John M. Olin of Madison. Now when I drive out of state visitors to view that natural beauty spot, they must also take in the scene of the shacks and rubbish that disfigure it.

I respectfully request that you inform me

1. Who authorized or is responsible for the mutilation of this property?

2. What state officer, if any, has authorized private parties to build shacks and take possession of and uglify this lake shore?

3. What steps, if any, can be taken by you to require the Superintendent of State Hospital, Dr. Green, to remove the shacks and the more offensive rubbish from the lake shore?

Kindly note that it costs no money to prevent the abuse of this property. It costs nothing not to dump rubbish or not to turn this State property over to private individuals for building of shacks and the exploitation of state property by private individuals. It may cost several days labor of the patients of the State Hospital to demolish the shacks and remove from the shore some of the most offensive rubbish. That is the total cost of correcting the mistake of not carrying out on its own property some of the State's teachings about conserving natural beauty.

But the difficulties were minor compared with the satisfactions; and although Mrs. Slichter sometimes would have preferred being at Frances Street for longer periods, she loved Sladshack and declared that the view across to Madison was superior to that of the Bay of Naples. Slichter's own opinion was given in a letter to Sumner on April 23, 1943, just after he became seventy-nine years of age: "Wish you were here to see the flowers at Sladshack. There is nothing so beautiful as a display of spring wild flowers. The beautifying of that bit of woodland is the best thing I have accomplished in my lifetime."

Family Transportation. I inquired of Louis Slichter how the family got back and forth between Frances Street and Sladshack; when his father got his first car; and whether they had horses. The reply (a portion of the letter of June 22, 1970) follows:

~ It is a special pleasure to answer those questions about the horse and buggy days. We had in succession three horses, within my memory, Darby, Mack, and Kit. The first two were driving horses for the two-seated surrey with the fringe on top. Darby was dark. Mack, whom I remember best, was a bright bay—light reddish. They, in succession, were kept at Brown's Livery Stable on State St—where Wehrmann's Leather Store now is (or was?), north side, just a few doors west of Gilman. The horse and buggy were delivered to our doorstep (WHAT SERVICE!) greatly to the delight of Bangle, the famous English pointer, and brother of Registrar Hiestand's dog. Bangle exuded his explosive enthusiasm by rapid gyrations of chasing his own tail in a tightly closed circle. Undoubtedly a dog enjoys an outing in the country —a picnic—more than any other creature because a dog invokes a whole new universe of smells which others completely miss. Well, we usually left the horse and buggy, or at least returned it to Brown's after leaving mother at 636, and walked the few blocks home.

In the fall, however, the horses were sent to pasture, on some farm, to reappear when the snow was gone. The third horse "Kit" belonged to a later era after we returned from abroad, and lived at the cottage except for a few months in the winter. . . . Kit was a western cow pony, and a riding horse, which I kept care of, and rode nearly daily— as did Mother and Father. I was in high school then. Bicycles for all were transportation to and from the cottage.

In winter transportation [in town] was generally by foot. . . . Street cars were always fun and a special treat.

It was in this era that Allen, Louis, and their father took a trip of a number of days on their bicycles, including visits to several Wisconsin towns and to Lake Geneva. It is still a cherished memory.

Slichter's first automobile was a 1913 Model T Ford.

The just quoted letter about their horses also says: "The only really miraculous annihilation of space which has personally struck me was the substitution of the motor car for the horse and buggy. That was when one suddenly found himself at new distant points quite inaccessible to a horse and buggy." One never remembers learning to crawl or to walk, which appears to be an even more liberating experience.

As indicated elsewhere in this chapter, Slichter grew fonder and fonder of exploring the countryside but his relation to a car was not one of complete symbiosis. I presume that we do not have records of all of Slichter's automobile mishaps. Madison tradition has it that Professor Edward A. Ross, Dr. Dorothy Reed Mendenhall, and perhaps Dean Harry L. Russell were worse drivers.

The last week of February 1915 must have been exciting on the roads. The following is from a letter of March 11, 1915, from the insurance adjuster concerning an accident which occurred on February 22:

~ *I met Mr. Harbort at Mr. Reed's office. His story is to the effect that you were absolutely on the wrong side of the road, that he saw you about forty rods away, that there was no snow or ice on the road, that he thought all the time you were going to turn out onto your proper side. He admits that there was nothing to prevent his giving way to you, but takes the position that because you were on the wrong side he had the right to stay where he was and that if he had moved he would have gone into the park with his bull which is against the law. This of course, is nonsense, as I told Mr. Reed. He says the stick he had was attached to the bull's nose, that it stuck out beyond his body and that the car struck the end of it. He claims that he was attempting to raise it up so that the car would clear it and that when it was about opposite his mouth it was struck and knocked against his teeth and he fell backwards in the road. He says the bull behaved like "a perfect gentleman" and had nothing to do with his fall. He has never had any doctor but claims he had a couple of teeth knocked out and says he went to a dentist a few days afterwards and does not know what the bill will be. I told Mr. Reed that I did not think the claim had any merit but I was always willing to buy my piece [sic] and that if his client was willing to take $50.00 I would pay the same. He stated he would talk it over with him and let me know.*

Slichter's account of the accident of February 28, 1915, was acknowledged by R. C. Nicodemus, insurance agent, on March 2 as follows:

~ *I beg to acknowledge receipt of your favor of the 1st inst. advising me that on the morning of Feb. 28th you ran into a young boy who jumped from a milk wagon onto the fender of the automobile; as you state he jumped up and said he was not hurt and ran along home I don't imagine anything serious will occur from it but if you hear anything further from him be kind enough to advise me. . . .*

There is also correspondence concerning Slichter's attempt to collect $20.25 from a resident of Lodi because of an accident which occurred on November 18, 1937.

There were other rather individual aspects to the Slichter driving. When he was old enough so that both his absentmindedness and his deafness had greatly increased, he would forget to take the car out of low gear, but would step on the accelerator and go along city thoroughfares at a maximum speed of twenty miles per hour, but with a decibel output equal to that of two unmuffled motorcycles, while blissfully unaware of the disturbance he was creating.

It was worse when he remembered to take the car out of low gear. A former secretary reported: "He drove like a maniac—passed street cars on State Street on either side as the mood struck him at as high a speed as he could hit."

Fauna. Children, horses, and dogs (Slichters are cynophiles) were not the complete menagerie. Chickens were present; and an annual pig, which seems to have been Allen's particular charge, was an important part of the household. His Aunt Belle remembered his high school graduation by five dollars sent from Europe to support "this year's pig." A cat adopted the family and brought her three kittens to share the hospitality.

The wildlife meant much to Slichter and his letters enumerated the birds he saw. Along with wild flowers, birds were among the favorite subjects—"hundreds" of bluebirds on the Sauk Prairie, the spring flight of warblers, the birds near the marsh at the base of Picnic Point, and ducks. This from a letter written in April, 1937:

This is a beautiful spring day—warm, calm, with birds everywhere. Yesterday, right in shore at my study windows was a flock of Butterballs—a kind of duck that used to be very common but is now so rare that I have not seen any for 40 years. In their spring Easter plumage they are very beautiful. They are called Butterballs because they are so spherical in form and so very fat. They are about the best duck for the table—hence their rapid extinction.

Each year he watched to see if a muskrat would reappear near Sladshack. He also watched for the loon. In 1942 an item in the *Capital Times* cast aspersions upon the bird. He replied:

~ Dear Times:
I disagree with the unfriendly words your Front Page put forth concerning the Great Northern Loon, one of the earliest spring tourists to visit the Wisconsin lakes and the best fisherman of them all. It is true that he shuns the cities and the factories and the smoke stacks that his fellow shipmate, Homo Sapiens, seems so proud of— but he is a friend of the lakes and the northland that Wisconsin boasts and boosts as vacation land. That call of the loon that you deprecate—is there any voice more expressive of the untamed North? Is that call ghostly and unhuman? Let us hope that it is! Perhaps the notes are poorly tuned and full of sadness, but they contain no threat

nor terror, nor do they herald calamity or destruction to others. It is only the far off call of the wilderness; it speaks of icy lakes, of the tundra and of the arctic isles, where he knows there is peace. Of all the creatures of the northland, the loon seems to typify best the stark wilderness of nature.

The terror and threat is not in the song of Gavia immer, but in the voice of his fellow traveller Homo sapiens, whose night cry is the roar of planes and bombs and the tumult of destroyed cities and the cries of torment from children and the defenceless.

Is the cry of Gavia immer music? I do not know. But sometimes in the great music of Sibelius I seem to hear the far off call of the lakes and the pines of the lonely North, all in the same mood as that night voice that comes from our Wisconsin lakes.

Gavia has found liberty, freedom and peace in his lonely life. Homo has sought these same things in his collective life, but as yet has not found them.

I must also dissent from your Front Page etymology. The word loon, meaning the diver, has no relation to the word loon, meaning a rogue or stupid fellow. The first word is from an old Norse root and may refer to the echos that loom through the heavy mists off the rocky shores of arctic lakes and fiords. It has nothing to do with lunatics, Luna, or the Moon.

A friend gave me a verse that I have attached to a copy of Audubon's drawing of the Loon:

> "A lonely lake, a lonely shore
> A lonely pine leaning on the moon;
> All night the water-beating wings
> And far off cry of mournful Loon."

Travel. After their first trip to Europe the members of the Slichter family never lost their taste for travel.

Slichter took great interest in the journeys of his children and his grandchildren, writing instructive letters to his grandsons. He particularly wanted Bill's reactions to the bust of the blind Homer in Naples. One of his last acts, less than a week before he died, was to offer to bear part of the expenses of Louis's trip abroad. We, of course, give no account of the extensive journeys made by the four sons. We only note that the second son was irate when, after the age of seventy, he was judged too old to participate in a geophysical experiment at the South Pole, although he had been a leader in the preparation for the work.

After the close of World War I, the elder Slichters traveled extensively. In the spring of 1924 he took a leave from the deanship (an example that the deans across the country might well follow) and went with Mrs. Slichter to Europe. The classicist, Professor Arthur G. Laird, acted as graduate dean during this period—to the greater order of the files. The trip was devoted to Italy, France, and England. He evidently did not like southern Italy or the evidences of Mussolini's dictatorship. John R. Commons wrote the minutes of Town and Gown for October 25, 1924, when Slichter reported on his trip, and concludes: "Altogether the T and G got a vivid picture of Slichter and wife doing just what they pleased and having a fine time in an effete civilization and glad to get back to U.S.A."

The Slichters made use of meetings as focal points of pleasure trips. Thus the 1933 meeting of the Association of American Universities took the Slichters by car to the Chicago World Fair, the Shenandoah Valley, Washington and Lee University, New York City, and Princeton, where the meeting was, and where he first met President Conant of Harvard, whom he describes as "Good hard-headed sane Yankee, not of family but of substance."

In 1936, now that he was no longer dean, he and Mrs. Slichter took a trip, the account of which is chronicled by Birge in the minutes of Town and Gown of March 28, 1936—a "twice told tale," which lost nothing on the second telling.

~ Happily for us, Slichter had returned to Madison in time for the meeting and he could talk of the 15 weeks trip while it was fresh in mind. Mrs. S. and he went from Madison to Cambridge via Washington, back thro' New York and Washington to Florida; an excursion to Cuba; then north via the TVA. He saw all of nature, man, politics, society and education within the area covered. Only mere hints can be given of his story. A little reticent on New York—"shows," he said. But what kind of shows?—! But for the rest—50¢ at Madison is worth $1.00 at Cambridge—Conant hard-boiled but efficient, will last a few years —winter resorts in Florida on a sand ridge 14 ft. above tide, between sea and swamp—nobody lives there thro' the year—Floridans can't cooperate, buyers fix price of citrus—soil lacks rare elements—western graft main value of ship-canal—Spartanburg (C.F.S.), Wofford College, coeducation not there, but classics—TVA, flood control, power, cotton mills and socialism—utlities selling "wall paper," not "juice"—all these and much more in profusion and full and pungently flavored.

So his story provided us with a "noble dish,"
                "A sort of soup or broth or brew,
                Or hotchpotch of all kinds of fishes—

> Green herbs, red peppers, mussels, saffern,
> Soles, onions, garlic, roach and dace."

All these he worked into "a rich and savory stew" and we, being "true philosophers," loved it for its "natural beauties."

In late winter and early spring of 1939 the Slichters took a motor trip to the west coast. In the same year Mrs. Slichter, escorted by the dean, made a pilgrimage to the homeland of the Byrnes. It was on this trip that he copied the testimonial to Monsignor Byrne. It was also on this trip that the charm of the young Irish women again struck him anything but speechless. On August 29 he wrote to Edwin B. Fred:

~ My dear Fred:

We have had a glorious time in our trip about Ireland. It is quite above expectations. In fact, when I return I am afraid I will be dubbed a liar on all points. Here is a sample—we have been here three weeks but have seen no rain—it has always been willing to rain when we were not about; not where we were. Trinity College is most impressive —the Library is, perhaps, its greatest glory. The interior is especially impressive. There is a fine large playing field which they say is crowded in term time but has a "stadium" seating capacity of about 100. The few students about are graduates and medics.

We have reservations on the Normandie sailing Sept. 6 from Southhampton. The uncertain war conditions make matters quite uncertain. If the French do not cancel the sailing we expect to get to Cambridge in time to help Conant start his University.

The Irish girls are simply superb. They can entertain wooden Indians. I asked one "Do you speak Gaelic"? She replied, "I never speak anything else." I said "But you seem to be speaking English now." "Yes," she said, "but you are the only man in the world I speak English to."

You can't beat them. With best wishes.

Perhaps the colleen had a Christopher for an ancestor.

This trip needed some pleasant relief, for while the Slichters were in Europe World War II broke out and they had some difficulty in getting passage home. They returned to Montreal in late September on the *Duchess of York* rather than to New York on the *Normandie*. This, I believe, was his last trip abroad.

In addition to wide-ranging travel, exploring the countryside around Madison was part of the way of life for the Slichter family. Picnics were often a portion of these excursions. We have already quoted the pre-engagement note in which Lulu Byrne offered to "hook" the lunch for their drive around Fourth Lake. When Slichter was in his late seventies he recalled early outings, for in 1941 he wrote: "You can skate all over the lake today. It is the first time in years that I wished to be young for a day, so I could take a skate sail, a necksack and food and go to the far parts of the lake for a fire and a picnic. It is glorious to see the others having fun, but I still think that they do not know as much about a complete winter venture on the lake as I did years ago."

Picnicking persisted. Beauty spots near Madison were used, the top of Blue Mounds, Colloday's Point on Lake Kegonsa, and one of the favorites—Parfrey's Glen. The frontispiece of this book is a photograph taken by Louis in 1940 during a picnic at our cottage near Okee.

But the lunch needed not be a picnic. On June 13, 1943, he wrote to Sumner: "Today we planned to take Mr. and Mrs. Guyer to Ft. Atkinson for dinner, but they persuaded us to eat at their house. It seems that they had a party last night and had lots of food left that they wished to get rid of, so we had to go. I greatly missed the trip through the beautiful Wisconsin country. There is no month like June and certainly there are not many Junes even in a long life time."

Such day trips had started many years earlier. For instance, in 1918 he describes his discovery of Cooksville, an experience which still delights the wanderer, even if Mr. Warner no longer lives to serve the knowledgeable guests.

~ Yesterday I drove about 32 miles to a place called Cooksville, near Evansville. It was a good town in early days before the railroad came through—it and a place called "Union" competing for trade and influence. Now it is nearly abandoned, but the old colonial style brick houses are left and one store and one blacksmith shop. A teacher of manual training in the Racine High School has purchased an old house and fitted it up with old furniture and utensils and lives there in the summer. He has also an old fashioned flower garden which is a blaze of glory. He serves meals to guests who write to him in advance. We had an old fashioned meal—ham and eggs, wild plum marmalade, wild gooseberry tarts. Coffee was served from a pewter pot and all the fittings on the table were old style. It was really quite interesting. He is just like a woman. He can weave, make rugs, do fancy work, knit, cook, put up preserves, do housework, etc.

For many a true traveler one of the joys is to stay home. Slichter wrote to Sumner on March 26, 1937:

~ Wednesday and Thursday were almost as good for me and almost as much fun as a trip to Europe. Tuesday night it started to snow and by Wednesday morning we were snowed in—in the afternoon drifts were as high as a car. Thursday morning it let up and the plows had cleared the road so we got our mail. It is equally delightful to know that you cannot travel as to know you can travel. I built up a big fire in my fireplace with huge logs while outside the beautiful storm raged. I had just started to write an address on Newton for the Autumn Meeting of the Math Society at State College Penn.—it is the 250th anniversary of the Principia. The storm was so delightful and stimulating that I got a lot done.

Family Correspondence. So many quotations from the Slichter family letters have been used that one might assume no further description of them is necessary, but they had certain characteristics that should be noted explicitly.

First of all the expressions of affection were warm and without self-consciousness. The salutations were "Dear Mom and Pop" or "Dear Mom and Dad" or "Sumner Darling" or "Dear 'fessor and Grandma Cakes." The great satisfaction of having affectionate sons or affectionate parents appears time and again, perhaps nowhere expressed more touchingly than in a letter to Louis. This is all the more so because it was written, not when Slichter was old and reminiscing, but just before his fifty-fifth birthday, when the sons were young men.

~ . . . This has been a proud day for me, for a copy of Sumner's book came. It is certainly a fine piece of work. I was deeply touched to note that he had dedicated it to his father and mother—for I tell you I care more for the affection of my boys than for anything else. I try to treat them square and to help them over places that seem rough at the time. It is always a surprise, however, to find that you boys notice things— your mention of the time your ice boat sank when you were home did me a lot of good, for it shows little things are not forgotten.

The above refers to the time Louis ran a new power iceboat over thin ice, where it broke through, and his father quietly had it raised that night.

He also wrote Louis, about the same time as the above letter, that Mrs. Slichter "certainly loves you boys as much as a mother dares. I often am very jealous of it."

Letters to and from the sons were frequent. Unless they were seeing their parents often, there would be apologies if more than a week or so elapsed between letters. Apology also followed when Louis's Christmas present for his mother had remained in the garage unmailed for ten months. Each letter in either direction brought joy.

The literary quality of the letters is high. Of course they are not labored. The punctuation is temperamental except that sometimes Mrs. Slichter put a period after every word. Spelling is casual. Words are repeated or omitted. The letters make clear how much working over the manuscripts of Slichter and his sons must have had prior to publication, but they also show how natural was the right word, the illuminating phrase, or the happy aphorism.

The topics were wide-ranging, but junkets, menus, and grandchildren were favorites. Family news was given in detail. Stories that one of the family had heard were often repeated. There were characteristic bits of humor; for example, from Sumner on August 12, 1944: "Today it reached 100. We should worry, however, because our car is protected to four below. Furthermore on Wednesday I bought a gallon of Prestone."

There was less discussion of University business or politics than I would have expected, although these are not entirely neglected.

A remarkable feature of these letters is the infrequency of disparaging remarks of any kind—infrequency, not absence—for the Kaiser, Hitler, and F.D.R. got their comeupances.

The Slichter sons, especially Sumner, had a most confusing habit of acquiring stationery from all sorts of sources—hotels, boats, business firms, rail and air lines, and army posts—and then using it for their personal correspondence without stating where the letter was written, while often dating it by the day of the week only. They had no consideration for the biographer who at first tried to reconstruct strange itineraries from these letterheads. Worst of all the letters sometimes actually were written at the places indicated. How was I to know that Sumner was in Boston when the stationery was from Harrod's, The Bohemian Club of San Francisco, the British Museum, or Sauk City Hotel and Tavern? Perhaps he was not. But though he used its stationery it was not he, but a McKinnon cousin that was in residence in the jail of Windham County, Vermont. Mrs. Slichter did not snitch, but she wrote her family on scrap paper, and Slichter often used the back of carbons or other outdated correspondence.

Some grand- or great-grandchild should edit or publish the Slichter letters. There are many gaps in the file but also much pay dirt. In this they should include the humorous sketches which he wrote for them. For instance, his grandson Charles received a fantasy entitled: "Good Taste—A Scientific Scoop." It concerns the Electrowalden Laboratory near Concord, Massachusetts, which developed complete diets of textured but tasteless pabulum which was to be accompanied by the desired taste on electro tastewaves. The development difficulties caused by taste-static from thunderstorms (accompanied by seasickness) or from electric razors were finally overcome. But you may never have heard of this discovery which will ultimately revolutionize civilization because the Laboratory "employs no Press Agent—in fact it employs a very able Hush Agent. . . ." Others than children enjoyed this flight of fancy for George Haight was given a copy, to his great pleasure.

Portraits. Slichter took great joy in the two portraits of his wife which he had had painted. These were done from photographs taken about the time of their marriage. The first portrait was painted by Harriett Blackstone, a New York artist with the attractive address, "The Garth of St. Mark's." This was done in 1930 as a surprise for Mrs. Slichter on their fortieth wedding anniversary. It is now the treasured possession of their grandson William. The second portrait was painted in 1936 by Benjamin Eggleston, who had already painted a portrait of the dean for the office of the Graduate School of the University of Wisconsin, as well as having made a pencil drawing of him which was used as the frontispiece of *Science in a Tavern.* The second portrait of Mrs. Slichter was given to their son Louis, and its painting was also a surprise to Mrs. Slichter.

After the second portrait was delivered, Louis wrote an ecstatic letter to his father: "The picture is a jewel, and throws a glamour over the entire dining room, just as mother would if she were there—I wish you both could see it." He then comments on the photographs used by the artist and continues, "I think it is very interesting that Eggleston did not think it advisable to change any detail of her dress, or adornments, for the purpose of artistic effect. The ear ring pendants, the lace collar, and brooch, even the buttons on the gown are all faithfully reproduced and the effect could not be more artistic. . . . Mary Lou, of course, will have it, since she, like me, carries the name."

To a Handful of Generosity Add a Pinch of Thrift. In Chapter 7 Slichter's business enterprises were described; but one's financial propensities are shown not only in how one makes money, but also in the way one spends it—both by what is bought and by the manner of buying.

He was particularly addicted to buying by mail and not chiefly from the great mail order houses. It was natural that many of his books came from Bowes and Bowes in London, but we do not know why some other purchases were not made locally. In 1892 he commissioned his brother-in-law George to get a baby carriage for Sumner in Chicago. His brother-in-law recommended a rattan one, which he described as "nobby and durable if not too fancy." Brother George at the same time bought him a basket for 75 cents that would "hold all the wood Louise wants to carry upstairs at once." George's letter ended prophetically, "We like the name of your boy and will honor it in days to come."

The subject of his mail inquiries ranged from seeds of rare plants to vacuum cleaners and newfangled starters for automobiles. He got oatmeal and tobacco of a special blend from S. S. Pierce of Boston. He inquired as to the price of wild rice from a man in Bayfield. (In 1940 it was 45 cents a pound.)

Moreover, he clearly took pleasure in reading the description of wares, especially hardwares, even when he did not buy. There are people like that!

Local buying was also carried on carefully. Even in 1891 the quotation he got of $27.00 could not have been an exorbitant charge for repairing the plaster; painting the trim; and papering the dining room, front room, back room, and hall.

When in 1936 he turned in a 1927 Franklin with a trade-in allowance toward a Ford of $75.00, plus any excess that it might bring, he got only the $75.00, since the new owner burned out the bearings because of an oil system failure after a few days of possession, and would not pay the dealer. In business as in sports, timing is important.

One amusing aspect connected with expenditures was that at times when he wanted cash he made the check out to "myself," and when he paid the bank for something, to "yourself," as shown in the illustration on the following page.

In Mrs. Slichter he had a ready ally. If the reply of the noble Scot to Samuel Johnson is correct, we have in the amount of oatmeal the family consumed an explanation of the success of her sons. But it was not only in feeding vast quantities of oatmeal to her family that she merited descent from the McKinnons. Sealskin coats were meant to last, for the one she owned before her marriage in 1890 was taken abroad with her in 1909 and registered with customs so that it could be brought back duty free in 1910. When, years later, her son Louis, who was passing through Chicago, called her by long-distance phone, she seemed to have been at least as disturbed by his extravagance as gladdened by his attention. Her ability to stretch recipes is still remembered with affectionate and amused admiration by her daughters-in-law.

The Slichters were generous indeed to their children, their grandchildren, and other causes they considered worthy. But in small mat-

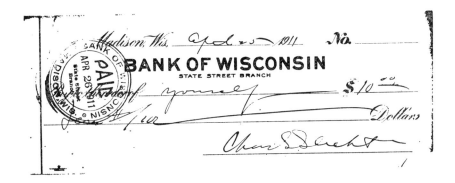

ters the husband wanted a good cigar for his nickel and the wife would rather have kept the nickel.

1910 to 1934

After coming back from Europe, the entire family was together in Madison for another four years.

In 1910 it was the father's hope that he soon could send the sons abroad again to study, and that he, himself, might make another extended visit to Europe. War intervened before this was possible (except for a summer walking trip of Sumner's) and led Louis to go abroad under very different circumstances.

We have little record of the family life for the period from their return from Europe until the United States entered World War I, that is, the period when the youngest grew from ten until he was almost ready for college, and the oldest entered college and almost completed his requirements for the Ph.D. Since Slichter's work on irrigation was tapering off, most of the family were at home. Sladshack was even more its base than Frances Street. The boys were old enough to sail, and evidently this was among their chief delights, for soon they were participating not only in local, but also out-of-town regattas. It was not long after the first automobile was purchased that some of them could drive.

When the United States entered World War I, the ages of the four sons were 25+, 21−, 19+, and 17−. Sumner finished his work for the Ph.D., and in June of 1918 entered the Army—less than two weeks after his marriage. He became a second lieutenant in December of the same year. Allen was an ensign in the Navy during 1918–19. By far the most exciting war work was that of Louis. Here Slichter's activities as a father and as a promoter of research merged. In this regard his actions of an official nature are described in Chapter 6.

The following letters of Mrs. Slichter to her sons Louis and Sumner are given in their entirety, both because of the light they shed on her special qualities and also because of the feeling she shared with other mothers with sons in military service. The common may be as important as the distinctive. She clearly believed that feminine concerns would be and should be of interest to her sons. The letter to Louis was written on Sunday, June 9, 1918—three days after Sumner's marriage. The letter to Sumner was also written in 1918.

~ Louis dear,
    This has certainly been a week for the Slichter family. I feel as if I had been drawn and quartered.
    The first thing of importance which you wish to know about is the wedding, of course. Your father and I went down on the six o'clock train, arriving about ten. We had to change our apparel so went to a hotel on 53rd Street—then took a taxi over to the Pence's just a couple of blocks away.
    The wedding was very simple, but very pretty. Ada looked very nice in her white dress of Georgette crepe and taffeta and her bridal veil. Helen, the one bridesmaid, wore a pink organdie. All the floral decorations were in pink—pink peonies in the living room and a large bowl of pink June pinks on the dining room table.
    The affair seemed to be in the hands of a colored caterer who answered the door and served the lunch which consisted of chicken salad, rolls, coffee, nuts, olives, ice cream and cake. We ate around the dining room. Allen came too late for the ceremony but not for the "eats."
    They had some very pretty presents. The Trumbowers gave them a Sheffield meat platter. Aunt Mide, a silver coffee pot. The Britting-

hams, a large silver basket for flowers; the Sharps, a china plate; (I am just mentioning the people you know who sent gifts) the Harpers, a silver sandwich plate.

We came back that evening on the 5:30 train, somewhat weary to say the least.

Friday morning I took Margaret over to get the cottage cleaned and we put in a long day at that. Then that evening there was Don's commencement. Don had the presentation of the class memorial and the introduction of Dean Birge, both of which he did very well. Before Mr. Miller presented the diplomas he gave a short talk and pronounced quite a volume of praise upon the class president, much to Don's embarrassment.

Sumner wrote that he and Ada would come on the noon train Sat. going right to Mendota so of course that necessitated someone's taking supplies over. I went over and had their lunch on the table for them but they didn't appear. We found a postal card upon arriving home saying they would be up at 5:30 and would come to the house for dinner, so then we hustled around to make preparation for it, and about 4:30 a telegram announced they would get here at 8:30, which they did. Don had purposely waited over to see them, but he was taking a train at 9:30 so his visit was brief. We take him to the train and them to the shack. They are coming in for dinner today. Al is here too and I have invited Aunt B. and Aunt A. I am sorry you will not be here to partake of chicken, asparagus from the garden, ice cream, and cake with frosting on it. Perhaps Sumner will be able to take you a box of cake when he leaves on Sat. otherwise I'll send you some later when I get over the rush.

Your father and I are going out to the Leith's for a picnic supper tonight. I think I shall have to have the next meeting at the cottage. I haven't had one in ages.

Your letter, telegram, flowers, and gift pleased Ada and Sumner immensely. You certainly have shown your affection for your brother and Sumner feels very happy about it.

With lots of love
Mother

Sunday, 10:00 a.m. August 18

~ Sumner darling,

The dishwashing seemed to take an endless time this morning. I can just write "finis." The trouble was I had let them accumulate for several meals and after awhile they do mount up even with two people.

Your father has just returned from the garden with a large basket

of tomatoes, a still larger one—a big market basket in fact—filled with
lima beans and about a dozen ears of corn. What a pity you and Ada
can't have some of the supply! Your father wondered about sending
some lima beans to you, but I am afraid it wouldn't be practicable to.

I am going to make a large dish of succotash for the picnic supper
this evening. The meeting is to be here at the cottage and it ought to
be fine with the lovely moonlight we are now having. We are going to
have SOME VIANDS, I tell you. I am baking some ham and Mrs. B.
[Brittingham] is going to bring peach ice cream. Then there will be
creamed potatoes, salad, and little cakes. I have some very delicious red
raspberry and black currant jam I made a day or two ago. I think I did
a foxy thing. I made a lot of apple jelly and just mixed with it about
one-third of the jam, so we have the quantity with sufficient of the
flavor, but not at all seedy. I hope your father will be able to take you
and Ada a sample at least.

I'd better go over the events of the week so you'll know what we
have been doing. I wrote Louis on Thursday and possibly he has sent
my letter to you.

We went to town Monday morning. When we left it was delightful, but it soon began to get hot and the afternoon was simply terrific,
also the next day until evening when a hard thunderstorm cooled
things off. We had planned to have a picnic out on the Point with the
Havens and Trumbowers but it was too hot to be considered.

We were in town all day Wednesday. I had a nine-thirty appointment at Dr. Hart's, but of course I did not get away until nearly noon.
I am still disabled from the treatment.

I had lunch that day with Agnes Keenan and later in the afternoon
we went out to the B's [presumably the Brittingham's] for supper.

I missed two opportunities of a good square meal because of the
soreness of my mouth.

The B place is a perfect dream. So many lovely flowers and the
sunset that evening was surpassingly lovely.

Thursday was a beautiful day and in the afternoon Mrs. Winegar,
Mrs. Stoddard—an old friend of mine who is visiting her aunt Mrs.
Winegar, Miss Agnes Keenan, and Miss Klauber came over on the two
o'clock boat. Mr. Winegar came on the train and your father met him
at the station.

They thought we had a fine supper and things did taste good and
certainly disappeared like magic. We had just quantities of lima beans,
fried potatoes, liver sausage, cheese, fresh bread, coffee, peaches and an
apple pie Mrs. Winegar made. I hardly knew the taste of pie, so long
since I had had any.

Friday was a day of rain and I was by myself all day. Your father was in town. I enjoyed the day very much however. There were all the papers to read besides magazines and then too I had knitting. Just before supper Mrs. Haven telephoned they were lonely and would we have supper with them. Of course it was a pleasure to accomodate her and we spent the evening in front of their grate fire.

Yesterday morning I went to the Capitol by invitation and worked for three and a half hours directing envelopes for the food administration. You will smile at my writing addresses, but I assure you I wrote much better than usual and took great pains to have my writing legible.

I was invited to have lunch with the Harpers and let me think—what news did I get? Well, the Marches have bought the W. V. Moore place out at the Heights, and it seems the Guyers were keen for the place too and made the Marches an offer, but they wouldn't consider it. [The Harper sisters were a great source of news not unmixed with editorializing.]

I think your father has sent you Don's letter which came this week. Don enclosed a letter Louis had written him which we enjoyed very much. It certainly is mighty funny. I read it to the Haven girls and they thought it was great. I'll send it to you in a few days. I think it will furnish the literary part of our program this evening.

Al too wrote several letters. His "awful cruise" must be about at the end. I am glad Ada is thinking of remaining with you. We are perfectly willing to help you, Sumner dear, and I don't think you should hesitate about accepting it. These are unusual times you must remember.

I must tell you what Miss Miller the nurse in Dr. H.'s office said to me on Wed. She said "I think you have the loveliest daughter-in-law. I took care of her grandmother and know them all very well." I agreed with her entirely.

You mustn't plan on my coming with your father. For many reasons I think he should go alone. It would mean much more of a vacation for him.

*With lots of love to you, Ada and Louis
Mother*

A postscript to one of Mrs. Slichter's letters to Louis written about the same time as above expressed at least curiosity: "We are all interested in the girls who will accompany your men to church. They must be quite a different type from any we have in the west. As far back as my time, that was a thing unknown, but perhaps the girl of the present day is of a more religious turn."

In comparison with Mrs. Slichter's view of Sumner's wedding we also have her husband's description, written to Louis on June 7, 1918.

~ We got back from the wedding last night. It was a very pretty wedding. Only immediate relatives were present—perhaps about twenty altogether. Rev. Hunt of Madison performed the ceremony and Ada looked lovely. Your mother had a new white dress, the material of which I got in California. She was by far the swellest person there. She certainly looked sweet and what if a tear did come to her eye. She had a right to it if anyone had. She has certainly been a fine mother to you all. Al got there a little late, but in plenty of time for the eats, which were served at about 12:30. Your mother has some cake for you.

It was during the period from World War I to Slichter's retirement, or soon thereafter, that the four sons became established in their careers. It was his privilege to be able to help his four sons substantially in the early stages of their development. He had the reward every parent craves, to see his children become public-spirited citizens, successful in their own endeavors, and good parents.

It is fitting in light of his father's later explanation to the grandchildren of the meaning of the name Slichter, that Louis, while still in high school, sold K-W Road Smoothers. Although the actual mechanical work was done elsewhere, the Frances Street home was the Smoothers' headquarters. We have no indication that this pleased Mrs. Slichter.

Retirement

During the twelve years of his retirement, Mr. and Mrs. Slichter nominally lived alone either at Frances Street or Sladshack; that is, no other member of the family was regularly domiciled at either place. Donald and Allen were living in Milwaukee; Sumner was in Cambridge; and Louis was either in Cambridge or Southern California, except for one brief period at the University of Wisconsin. Sumner spent a number of his summers in his cottage adjacent to Sladshack. All the sons and their children frequently visited the parents, to the joy of all concerned.

The Slichters adjusted well to retirement. In 1936 Slichter wrote to his son Louis: "Your mother and I are very happy and we are having a fine time in our simple, unpretentious way. It would be ideal if every-

one could have as happy a life as we have had." The remaining decade
was to prove equally happy.

The years of retirement could, as well, be called "The Years of the
Grandchildren." Slichter concluded his "Two Hundred Years of Pioneering" as follows:

~   *After two hundred years of pioneering, and of plowing and replow-
ing the soil, the family tree is still sturdy and blossoms as never before.
The Swiss stock, cross fertilized with other hardy stock, now bears a
luscious harvest of*

Peaches on the Family Tree
   (1) *Sumner Huber and Ada Pence Slichter had two children:*
       William Pence Slichter born March 30, 1922
          married June. . . . .
       Charles Pence Slichter born January 21, 1924
          married June. . . . .
   (2) *Louis Byrne and Martha Buell Slichter had two children:*
       Susan Merry Slichter born January 15, 1928
          married June. . . . .
       Mary Louise Slichter born December 26, 1929
          married June. . . . .
   (3) *Allen McKinnon and Dorothy Fritsch Slichter had two children:*
       Marjorie Ann Slichter born November 27, 1923
          married June. . . . .
       Donald Allen Slichter born February 18, 1932
          married June. . . . .
   (4) *Donald Charles and Dorothy Doyon Slichter had two children:*
       Ann Slichter born January 10, 1932
          married June. . . . .
       Jane Slichter born October 24, 1935
          married June. . . . .
*The job of a Slichter, if true to his name, is to make smooth for
others the rough places along their way.*

How could Slichter expect grandchildren, particularly his grandchildren, all to be such conformists as to marry in June? In fact of seven
who have married, only Sumner's son William was married in June, and
in reply to my threat he wrote: "You are welcome to blame my June
wedding on my wife. You may also want to blame the academic community which lets students out of jail in June."

Seldom have greater or less successful efforts been made to spoil
grandchildren. Thus his father wrote to Sumner the day after Christmas

of 1932 concerning Allen's son: "Little Don contributed a lot as he looks just like Sumner did at his age and it was as if 636 of 40 years ago had come back again. He is as jolly and bright as he can be. Just like Sumner was, for I can teach him any trick in just a few minutes. He has a good chance, of course, of being spoiled by his Grandparents, but that is a chance most children have to take."

During these retirement years Sumner's sons grew to manhood and entered military service; Allen's daughter became a young woman; Louis's daughters reached their teens; and Allen's son and Donald's daughters were small children. Large portions of the letters between the grandparents and their sons were devoted to the grandchildren and there was considerable correspondence directly with the grandchildren.

Sumner was justly proud of the records his sons made at prep school and at Harvard, and knew his father would be delighted.

The following letter shows the close relationship between Charles Pence Slichter and his grandparents, and it also shows that the sense of humor was passed along to another generation. We believe it was written while Charles was an underclassman at Harvard.

Monday

~ Dear Grandma Cakes,

This is just a very short note to warn you of impending disaster— I'm taking you up on your invitation to come to Madison. You're really awfully swell to ask me! I guess Bill isn't able to make it; however, I certainly can and want to come!

Tomorrow I finish exams and move home, so I plan to leave here on Thursday. The New England States, which I plan to take, will get me to Chicago next morning. I'm going up to Evanston to see my Grandfather Pence and Aunt Nelle, so I'll come up to Madison Friday evening by way of Milwaukee. I'm not sure what time the train arrives, but I'll send you a wire from Chicago.

Bill, the dog, has been keeping secrets from me! He just informed me the other day that he has some more Defense stamps you sent us. So please excuse my delay in thanking you. It was wonderful of you to send them and I thank you loads even though I'm pretty late in doing so!

I must stop now because I have to get to work learning German words! We had a rather gruelling physics final this morning, but I don't think I did too poorly, although I did far from well. On glancing over this letter it looks pretty feeble, so please blame it on exams. I'm so shot I can't seem to express how darn excited I feel about going to Madison!! You're swell to ask me! Give my love to Fessor and warn

him, too, so you'll both get a chance to hide! See you both Friday
night!

> *Lots of love*
> *Charles*

Sumner's sons originated the titles "Grandma Cakes" and " 'Fessor,"
and the younger cousins adopted them. The wives of her sons first
wrote to their mother-in-law as "Mrs. Slichter" but soon succumbed to
"Grandma Cakes."

It was a still more youthful Charles who, after inquiring about the
salaries of monks, decided to be a professor so as to have summer vacations. He has his wish.

Nor does the new generation lack the power of vivid description.
Sumner quoted to his parents Bill's letter from the Philippines, written
in 1943:

~ *If you have any thought of coming over to this part of the world*
some day on a rubberneck tour, ye gods, give it up right now! Certain
parts of Manila might have been beautiful once. On my way to the
baggage warehouse I hiked through the filthiest, smelliest place that I'd
ever seen. Incidentally it was one of the main streets. The streets them-
selves were a juicy, black mass of mud, littered with garbage, refuse,
papers, tin cans, pigs, dead dogs, and other pleasant things. The stench
of the whole region simply defies description.

Each stage in the growth of the grandchildren was reported to the
grandparents and re-reported by them to the rest of the family—Ann's
first outdoor walk, little Donald's first tooth at about two months old,
Mary's learning to swim or to ride horseback. Even Dort, the dog of
Louis's family, sent a kiss through Donald back to 'Fessor.

Amusing sayings were also reported. "You know, Mummy," said
Mary Lou, "I liked that red coat of mine so much that I always used to
think it would grow bigger with me and fit, but it didn't." And, "Ann
says that Grandma Cakes talks 'sensible,' but that 'fessor talks 'crazy.' "

When in 1938 Sue went to camp she received the following letter.
The envelope read:

> *Professor Susan Slichter*
> *Camp Kehonka*
> *Wolfeboro*
> *New Hampshire*

Just one door east   )
)
of their west neighbor.)

And the letter:

~ *My dear Susan:*
  Well, it was fine to get your good letter. It must be a lot of fun at
camp. We are having a good time with Bill and Charles. They go
sailing every day. Last Saturday we went to a cocktail party at Little
Norway near Blue Mounds. It was really not a party because they did
not have ice cream and cake. They had rolls and spiced meat and
caviar (it comes off of a fish) and anchovies and three kinds of cheese,
two that came off of a cow and one kind that came off of a goat.
  There are a lot of wild animals at Little Norway, but not the kind
that hurt you. There are partridge, quail, brook trout and pheasant. I
usually do not spell fessant that way but today I do. We have two ten
pound land locked salmon sent to us from New Brunswick, which we
must eat before they get too old.

Much love to you,
Fessor

What the grandparents meant to the grandchildren can be seen
from two letters, one written to Mrs. Slichter by Charles immediately
after her husband's death, and the second written for me in December
of 1968 by Donald's daughter, Ann Pike.

~ *Dear Grandma Cakes,*
  We were all so sorry to hear the sad news about Fessor, and wish
we could be out at Madison at this time. It was particularly nice for
Bill and me to have been able to have had our visits in Madison this
summer after the interruptions caused by the war.
  So many of my friends haven't known their grandfathers. I always
think how lucky Bill and I were to be able to spend our entire summers
near Evanston and next door to you and Fessor. A lot of students
considered themselves very lucky to have been associated with Fessor,
but to have such a wonderful man for your own grandfather gives you
a special thrill of pride! It takes a very patient man to let a couple of
kids such as Bill and me run wild in his workshop for several months
every year, but it certainly was good for us!
  Bill and I have another advantage over our friends besides having
known Fessor so well; we have a very high example of a man and a
scientist to live up to. It makes me want to do my best because of Fes-
sor's approval whenever I've had some measure of success. During the
war when lots of people looked askance at a guy my age not being in
uniform, I always felt I had a good solid backing of some people who
were proud that I could do something in science for the war effort. In

the same way, as an officer Bill had a wonderful example to look to in fairness and taking an interest in those under him. A fellow has a great advantage if he has a concrete ideal to live up to, because he sees that someone else could reach that goal!

We are thinking of you all so much and wish we could be on hand. It's so nice that Uncle Don, Uncle Allen, Uncle Louis, and daddy are in Madison.

Lots of love
Charles

And, from Ann Pike:

~   We all called my Grandfather, "Fessor." A short cut to Professor!

I loved being with him because of his warmth, gentleness and sense of humor. His wonderful interest, curiosity and knowledge of his world around him made it come very much alive to us as children. The beauty of his wildflowers in the woods, the birds he fed outside the window from suet hung from branches, the fresh vegetables he grew, his books, his woods, the lake. This was Fessor's world when we were children— the one we shared in and loved also.

I was very shy as [a] child but Fessor had a wonderful way of making even a very little girl feel rather big and important. He more than any-one else in my family took a genuine interest in mine, which was art. My most cherished memory was of his sitting down with me as I drew a picture, and drawing a wonderful paper man, one that he used to make as a school boy. When the paper used was folded properly, this clever paper man would stick out his tongue! Not only was I amused and delighted but promptly upon taking that wonderful, saucy paper man home, inscribed [it] for posterity, made by Fessor on August 23, 1941. This I knew I would always cherish. I was 9. He was 77.

My final pet memory of Fessor, was the way in which he would often and spontaneously in the midst of whatever was taking place, lean over and squeeze my Grandmother's hand, look at her with a soft twinkle and say "Dear Lou," that was all. His affection for his family was so unselfconscious and we loved him dearly.

But it would be a great mistake to envisage Mr. and Mrs. Slichter, after his retirement in 1934, as two old people basking in the light of happy memories, enjoying the successes of their sons, and delighting in the occasional company of their grandchildren. Besides taking pleasure in these occupations when they had time, and in driving around Dane and its contiguous counties, they were busy folks. The dean's retirement was only partial for, upon the invitation of his successor—Dean, later to Slichter's delight President, Edwin B. Fred—Slichter kept a desk in the Graduate Office (which the secretaries learned not to touch), and had a part-time appointment as a consultant on research. In addition, it was after 1934 that he edited *Science in a Tavern* and wrote several of his essays. It was six years after retirement that he started helping small colleges by nest-egg donations of interesting and valuable books. The studying of the colleges, the purchasing of the books, and the describing of their importance to the recipients was a labor, even if a labor of love. He read much and his sons frequently gave him books for Christmas and birthdays. For Mrs. Slichter, as in the case of many other wives, her husband's retirement could mean an increase, rather than a surcease, of work.

Of course, as the years went on, Slichter lost some of his physical energy but he did not lose his clarity of mind nor his zest, and as ever, he liked young people. One month before his death he wrote to his son Louis:

~ *I was so pleased to get your good letters. Sumner left Saturday and Don and family left early this morning—6:30. They were out until 5:00 in a launch belonging to Mr. Vilas of NWRR and the girls had a great time enjoying the race. One of the Vilas girls sailed their cat boat. The wind was low at noon but came up at 3 o'clock. At five o'clock they went ashore and the Vilas boy and girl got them all a fine picnic supper. The kids have been trained to do it all by themselves—the boy in the kitchen and the girl in the dining-room. The children have all been trained to cook. That boy will make a fine mate.*

I am so glad you can stop off. This is my 60 anniversary of entrance into Madison, my lucky day! Ann and Jane insist that Don move to Madison. They are crazy about sailing. They sailed this summer in large cub boats.

Much love to all
Pop

Soon after this letter was written, Slichter went to the University of Wisconsin Hospital because of a heart ailment and died there on the evening of Friday, October 4, 1946.

Once during these last days his doctor, Frank Weston, entered the hospital room, saw the elderly couple holding hands, and heard Slichter say: "Lou, you are beautiful!" The doctor left unobserved but still cannot tell of this occasion without choking.

Mrs. Slichter survived her husband by nearly nine years, continuing to live at the Frances Street home and at Sladshack, visited by sons and grandchildren. As she had graced youth, so she graced old age—capable, intelligent, kindly, serene, and rich in memories.

After a simple funeral at Grace Church, Charles Sumner Slichter was buried in Forest Hill Cemetery in Madison. Through golden sunshine, the ripened leaves of the oaks, without haste, were falling to the ground, covering it with their autumn glory.

CHRONOLOGY ❧ INDEX

Chronology of Charles Sumner Slichter

1864	Born April 16, Saint Paul, Minnesota
1869	Moved to Chicago, Illinois
1873	Spent summer with relatives in Ontario, Canada
1881	Entered Northwestern University, Evanston, Illinois
1884	First scientific publication
1885	Bachelor of Science, Northwestern University
1885–1886	Instructor at Athenaeum, Chicago, Illinois
1886	Instructor, University of Wisconsin, Madison, Wisconsin
1889–1905	On the Athletic Committee in its various incarnations, most of the time as chairman
1889	Promoted to Assistant Professor
1890	December 23, married Mary Louise Byrne
1892	Promoted to Professor
1892	January 8, son Sumner born
1892	Land for Sladshack acquired
1894	Started work with Professor King on underground waters
1896	May 19, son Louis born
1897–1899	Member of Common Council, Madison, Wisconsin
1898	February 18, son Allen born
1899	Publication of "Theoretical Investigation of the Motion of Ground Waters"
1900	President, Wisconsin Academy of Sciences, Arts and Letters
1900	July 3, son Donald born
1901	Joined Madison Literary Club
1902	Joined Town and Gown
1904–1905	Surveys of water resources near Garden City, Kansas
1905	Construction of Pumping Station at Garden City authorized
1905	Central Realty Company formed (Carroll Street Building)

1906   Construction near Garden City, Kansas
1906   Appointed Chairman of Mathematics Department, University
            of Wisconsin
1906   Actively leading in establishment of University Club
1909–1910   Year Abroad
1909   "The Rotation-Period of a Heterogeneous Spheroid"
1911   "The Mixing Effect of Surface Waters"
1912   Report on water supply of Winnipeg
1915–1916   Associate Editor, *The American Mathematical Monthly*
1916   Honorary Sc.D.—Northwestern University
1917   Appointed chairman of newly formed Research Committee;
            remained chairman until retirement in 1934
1917   Active in supporting research on submarine detection
1920   Appointed Dean of Graduate School, University of Wisconsin
1924   Trip to Europe
1925   August 5, Regents resolution against acceptance of gifts from
            Foundations
1925   Organization of Wisconsin Alumni Research Foundation
            (Articles of Organization filed November 14)
1928   First grant-in-aid of WARF (E. B. Fred and W. H. Peterson)
1929   First two regular WARF grants on recommendation of Re-
            search Committee: (1) Computer Service, (2) Professor
            Roebuck
1930   Regent action on gifts rescinded
1932–1933   First WARF postdoctoral grants
1933–1934   Grant for research appointments to faculty from WARF (in
            part from capital)
1934   Retired from deanship and professorship; appointed as special
            research advisor in the Graduate School
1938   Published *Science in a Tavern*
1939   Trip to Europe, including Ireland
1942   November 17, "Galileo"
1946   October 4, died, Madison, Wisconsin

Index

Accrediting of High Schools and Appointments, Committee on: CSS member of, 127

Adams, Charles K.: letter to Regents on athletic matters, 108; requests CSS to speak at Convocation banquet, 137; administration of, 147; member Town and Gown, 214

Adams, Thomas S.: replaces CSS on Athletic Council, 105

Adams Hall: men's dormitory, 116, 117

Adkins, Homer: organic chemist, 162

Advanced degrees: number of, 157

Agriculture, College of: CSS praises, 173

Alcohol: from beets and potatoes, 81; tax on, 81

Allen, C. E.: objections to German visiting professors, 169

Allen, Florence E., 40

Allen, William H.: survey of University of Wisconsin, 128–29

Allmendinger, David: analysis of instructional reports of CSS, 22; interviews with Otto L. Kowalke, 30, 116; interview with Karl Paul Link, 161

Alsted, Lewis L.: letter from CSS on alumni interest in athletics, 110

American Mathematical Society: CSS toastmaster at 1927 dinner, 29; name changed from New York Mathematical Society, 39; mentioned, 24, 31, 41

American Telephone and Telegraph Company, 31

Amish, 8

Anderson, C. J., 187

Anecdotes: CSS as teacher, 32–34; CSS as Graduate Dean, 150–52

Anti-Saloon League: complaints about, 197–98

Applied mathematics: CSS's working knowledge of, 22; textbooks, 38; at University of Wisconsin, 52

Arizona, University of: CSS considered for presidency, 172

Arkansas River: Garden City located on, 70; Owen's letter on flooding of, 71; measurement of underflow, 76; dispute between Kansas and Colorado over use of water from, 81

Arkansas valley; crude oil used as fuel, 74, 75; cost of pumping water, 90

Association of American Universities: 170; speech given on November 10, 1927, 170–72

Atchison, Topeka and Santa Fe Railroad: Garden City on main line of, 70; letters from CSS to Wesley Merritt, Industrial Commissioner of, 72–75, 91; complaint from CSS, 82

Athenaeum, Chicago: CSS taught at, 20

Athletics: Committee on, 103–12; Athletic Council, 105, 108, 109; intercollegiate, professionalism in, 106–8;

Athletic Board, 108; alumni interest in, 109; financial problems, 110; football coach hired by Athletic Association, 109, 111; salary of coaches, 111; rules governing personal conduct of athletes, 111; Athletic Council approves baseball schedule, 112

Automobile mishaps, 273–74

Babcock, Rodney W., 45

Babcock, Stephen M.: milk test, 57, 68; CSS gives mathematical assistance to, 68, 69

Banting, Frederick G.: fund from development of insulin, 178

Bardeen, Charles R. (Dean of School of Medicine), 136, 148–49

Barnard Hall: women's dormitory, 114

Barnes, Charles R.: member of first Athletic Council, 105; member Town and Gown, 214

Bascom, John, 25, 145, 147

Bascom Hall (originally called University Hall), 47

Baseball, intercollegiate, 111–12

Bashford, James B.: name suggested for dormitory, 118

Beardsley, Edward H.: biography of Harry L. Russell, 148

Beasley, Howard, 33

Beasley Tract, 9

Bechtel, Elizabeth (wife of Jacob Bechtel), 11

Bechtel, Elizabeth (wife of John Schlichter), 9, 11. See also Slichter, Elizabeth (Bechtel)

Bechtel, Jacob (father of Elizabeth Bechtel Schlichter), 11

Bell, W. M.: secretary, Finney County Farmers' Irrigation Association, 76–77

Bell Telephone Laboratories, 31

Berlin, Ontario, 14

Berne, Canton (Switzerland), 11

Birge, Edward A.: comment on, by Sidney D. Townley, 29; decision to appoint Van Vleck to professorship, 43; quip to George C. Sellery, 89; head of Library Hall Committee, 127; defense of University's inspection of high schools, 128; announces annual legislative appropriation for research, 142; nominates CSS as dean of Graduate School, 143–44; letter from CSS on requirements for faculty, 144; honored for 40 years of service, 145; educational background, 145; secures increased standard of pay for faculty, 146; tribute to, from Charles H. Haskins, 146; salary as president, 147; supports resumption of Carl Schurz professorship, 170; request from Steenbock for half-time assistant, 179; member Town and Gown, 214, 220–21, 277–78; mentioned, 22, 102, 111, 132–35 passim, 145–47 passim

Blackstone, Harriett: painted portrait of Mrs. CSS, 282

Bliss, Gilbert A.: considered for professorship, 42, 43

Bocher, Maxime, 41

Botkin, Alexander C.: name suggested for dormitory, 117

Bowman, Isaiah: given information by CSS on water rights, 100

Bradley, Harold C.: chairman of Committee on Undergraduate Social Needs, 116; correspondence with CSS on Experimental College, 119–22; letter from CSS on Memorial Union Building, 123–25; mentioned, 102

Brink, R. Alexander, 187

Brittingham, Thomas E., 267

Brittingham, Thomas E., Jr.: letter from CSS asking for funds to support Karl P. Link's work, 159–61; Articles of Organization of WARF, 182; investment of income from patents, 184

Brooklyn, New York: flow of underground water, 63; water supply, 92

Brown, Timothy: Articles of Organization of WARF, 182

Brown City, Michigan, 12

Buck, Philo: member CSS memorial committee, 247

Bunn, Romanzo, 145

Burd, Mr. (U. W. treasurer's office), 109

Burgess, Horace T., 45

Burke, Laurence C.: associate librarian, University of Wisconsin, 166, 267

Business activities of CSS, 194–204

Butler, Harry: articles of incorporation and bylaws of WARF, 176; charter of WARF, 182

Butts, Porter: CSS responsible for men's halls being house units, 116; memorandum concerning CSS and Union building, 125

Byrne, Agnes (sister of Mrs. CSS), 267

Byrne, Bessie (niece of Mrs. CSS), 202

Byrne, Donald: mentioned in letter from Sumner Slichter to CSS, 79

Byrne, Ellen (sister of Mrs. CSS). See Merrill, Ellen (Byrne)

Byrne, Ellen Sheehan (grandmother of Mrs. CSS), 248

Byrne, George (brother of Mrs. CSS), 265, 283

Byrne, Isabelle (sister of Mrs. CSS), 209, 265–66

Byrne, John (brother of Mrs. CSS), 196

Byrne, John A. (father of Mrs. CSS), 248, 265, 266

Byrne, Maria (sister of Mrs. CSS). See O'Dell, Maria (Byrne)

Byrne, Maria McKinnon (mother of Mrs. CSS), 248

Byrne, Mary Louise: married CSS, 245; elementary teaching career of, 250–51. See also Slichter, Mary Louise (Byrne)

Byrne, Right Reverend Mgr. (uncle of Mrs. CSS), 248

California: underground water, 63; survey of water resources, 69

Callan, John G., 140

Calvinists, 8

Capital Times (Madison): letter from CSS protesting article about loons, 275–76

Carl Schurz Memorial Professorship, 169

Carnegie Institution, Washington, D.C.: publication of The Tidal and Other Problems, 64; grant from, 65

Carson, John: at AT&T, 31

Carstensen, Vernon: with Curti, History of the University, 103, 128

Cederberg, William E.: Ph.D. thesis under direction of CSS, 23

Central Realty Company: formed 1905, 196. See also Charles S. Slichter Company

Chadbourne, Paul A.: CSS paper on administration of, 227

Chadbourne Hall: women's dormitory, 114

Chamberlin, Thomas C.: relationship with CSS, 21, 65, 69; paper published by Carnegie Institution, Washington, D.C., 64; CSS comments on book about tidal problem, 66; introduction by CSS at Birge's dinner (1915), 133; CSS collaboration with, in geophysics and cosmology, 139; Birge complimented by CSS for contribution to success of administration of, 146; mentioned, 57

Chamberlin-Moulton planetesimal hypothesis, 65

Chanute, Kansas: carload lots of oil shipped from, 75

Charities, 239–40

Charles S. Slichter Company (successor to Central Realty Company), 196

Chase National Bank, staff of: contribution toward war research work of Max Mason and Louis Slichter, 141

Cheynowith, Lt. Edward: member of first Athletic Council, 105

Chicago, Illinois: CSS's family moved to, 11; CSS's boyhood in, 12; CSS account of trip to, with Louis and Allen, 256

Chlorination of water supplies, 98, 100. See also Winnipeg

Christopher, Henry: letter to Ellen Sheehan, 248–49

Christopher, Thomas: letter to Ellen Sheehan, 249

Cimarron River project: letter from CSS to Wesley Merritt, 91

City water supplies. See Water supplies, city

Civic activities of CSS, 204–12

Coffin, Victor: service on high school inspection teams, 127

Cole, Frank N.: secretary of American Mathematical Society, 41

Cole, James: member of first Committee on Athletics, 105

College Club: complimented by CSS, 124

Colleges of State, relations with: 167–68

Colorado: dispute with Kansas over use of water, 81

Common Council, Madison: service of CSS on, 205–8

Commons, John R.: appointed to University Research Committee, 140; late in reporting grades of graduate students, 166; minutes of Town and Gown, 277; mentioned, 102

Computer service: established as first in American universities, 186–88

Comstock, George C.: on Fry's examining committee, 32; suggestions from CSS on how to increase graduate student enrollment, 139; retires as Dean of Graduate School, 143; CSS considered possible successor to, 144

Conestoga wagon: invented in Lancaster County, Pa., 9

Conservation Commission, State: letter from CSS on pollution in Lake Mendota, 271

Constructional Development Committee: continued study by, of building for student housing, 114

Control, State Board of: letter from CSS to, on pollution in Lake Mendota, 271

Consulting projects of CSS, types of, 69

Cornell University: outstanding as graduate institution in agriculture, 173

Cottage on Lake Mendota, CSS's. See Sladshack

Cunliffe, J. W.: minutes Town and Gown, 264

Curti, Merle: with Carstensen, History of the University, 103, 128

Customs Collector, Boston: CSS protests charge by, 263–64

Dahl, Gerhard M.: contribution toward war research work of Max Mason and Louis Slichter, 141; active in establishment of WARF, 176

Daily Cardinal, The: CSS's connection with, 112–14; comment on interest in athletics, 112; faculty influence on, 112; relation to faculty and regents, 112; financial difficulties, 113

Daniels, Farrington, 187

Darcy's law for velocity of flow of liquid, 58, 60

Darwin, Charles: theory of evolution, comments on, by CSS, 19; comments on, by Frederic Slichter, 20

"Debunking the Master's Degree," 170–72

Deerfield, Kansas: pumping plant near, 75, 85; underflow of Arkansas River, 76; authorization for plant at, 79; mentioned, 83

Depression: effect on number of students, 154, 155, 157

Detweilers (Swiss-Canadian relatives of CSS), 14

Disorders, Committee to Investigate Recent (1889): CSS member of, 126

Doctoral degrees: under CSS, 23–24; number granted in mathematics before 1906, 41; method for raising standards for, 152; fee for publishing thesis, 154; number advanced degrees before and during CSS's deanship, 157

Doherty, H. L.: letter from CSS concerning University Club, 130

Dollard, John (secretary of Memorial Union Building Committee), 125, 126

Donnelly, Theo, 151

Doon, Ontario: mill of CSS's uncle at, 14

Dormitories: committee on, 114–22; cost per student, 115; naming of, 117–18

Dowling, Linnaeus W.: college texts, 195; mentioned, 39, 40, 44, 48

Drainage flume: interference with water level in wells, 59

Drake University: CSS official delegate to 25th anniversary of, 138

Dresden, Arnold: influenced Thornton Fry, 32; college texts by, 195; mentioned, 45

Duggar, Benjamin M.: research in viruses, 159

Eggleston, Benjamin: painted portrait of Mrs. CSS, 282

Elsom, James C.: member of first Athletic Council, 105

Engineer, CSS as, 69–101

Enrollments: in CSS's classes, 22–23; in mathematics department, 47; in Graduate School, 154–55; of foreign students in Graduate School, 157

Entrance Requirements, Committee on: CSS member of, 127

Europe: Slichter family's year in, 259–64

Evans, Evan A.: letter to CSS concerning graduate manager of athletics, 109; trustee of WARF, 110; suggests names for dormitories, 117; opposition to use of capital of WARF, 190; mentioned, 113

Evolution, theory of, 19–20

Experimental College, 119–21

Eyster, J. A. English: appointed to University Research Committee, 140

Faculty citizen: CSS as, 102–38

Fallows, Samuel: name suggested for dormitory, 117

Fairchild, Lucius: member Town and Gown, 214

Family life: of CSS, 248–95; of Canadian relatives, 12–14

Farley, J. H.: letter from CSS re athletics to, 110

Farkasch, Hazel (secretary to John R. Commons), 166

Farmers' Ditch: water supply increased in, 75; letter from CSS concerning, 76–77; trouble with users of water from, 83; in disrepair, 86; mentioned, 79, 81

Faville, Henry B.: name suggested for dormitory, 118; introduced by CSS at

dinner for E. A. Birge, 134; mentioned, 136

Fellowships: stipend raised, 155

"Field Measurements of the Rate of Movement of Underground Waters," 63. See also Water Supply and Irrigation Paper, No. 140

Finch, Vernor, 177

Finney County, Kansas: Garden City located in, 70

Finney County Farmers' Irrigation Association, 76

Finney County Water Users' Association: mortgages on land owned by farmers, 83; successor to Finney County Farmers' Irrigation Association, 85; liens against, 86–87

Follansbee, Robert, 62

Food and drink: the Slichters' preferences in, 267–70

Ford, Guy Stanton: advice from, on appointments to Mathematics department, 41, 42

Foreign students: scholarship funds for, 156; increased number in Graduate School, 157

Foundations: resolution of Regents to refuse gifts from, 175; resolution rescinded, 175, 176

Fowlkes, John Guy, 162

Frank, Glenn: called on to settle salary increase argument, 51; proposed letter to, from H. C. Bradley, 119–22; letter from CSS to, on fellowships for foreign students, 157; letter from CSS concerning Perry Wilson, 161; letter from CSS concerning cut in University Research Fund, 173–74; mentioned, 35

Frankenburger, David B.: name suggested for dormitory, 117; member Town and Gown, 214; mentioned, 145

Franklin County, Pa., 8, 11

Fred, Edwin B.: urged writing of biography of CSS, 4; letter from Rosetta Mackin to, 35; recommends Perry Wilson, 161; CSS refuses funds to buy reprints, 175; grant-in-aid for study of bacteria, 186; member CSS memorial committee, 247; letter from CSS in Ireland, 278; used CSS as consultant on research, 294–95; became president to CSS's delight, 294–95; mentioned, 34, 162

Freeman, John C., 145

Frontispiece: photograph of CSS taken by Louis used for, 279

Fry, Thornton C.: Ph.D. thesis under direction of CSS, 23; comment on CSS as teacher, 31–32; mentioned, 31, 33, 52

Gabriel, John H., 78

Gale, Henry G.: letter from CSS re Mason, 163

Galena, Illinois: family of CSS in, 11

Galileo: paper on, by CSS, 228–30

Garden City, Kansas: irrigation project, 69–88; use of small pumping plants, 74; cost of fuel for pumping plants in, 74–75; cost of water, 77; Deerfield project, 82; rainfall, 84; high cost of irrigation to individual farmers in area of, 86; expensive debacle, 87

Gisholt Machine Company: contribution toward war research work of Max Mason and Louis Slichter, 141

Golden Vector, The, 6

Gordon, Charles E. (resident engineer for Garden City–Deerfield project), 82–83

Grace Church: CSS's relationship to, 211–12; Mr. & Mrs. CSS members of, 250–51

Graduate deanship, CSS's years in, 139–93

Graduate students: financial problems of, 154

"Grandma Cakes": grandchildren's nickname for Mrs. CSS, 268, 292

"Greens Farm" (home of Christian Schlichter), 11

Gregory, Charles N.: member Town and Gown, 214

Gregory, Stephen S.: name suggested for dormitory, 117

Griffin, Frank L.: on teaching of mathematics, 38

Ground water: "Theoretical Investigation of the Motion of Ground Waters," 57–60; flow of, 60, 74, 95; motion affected by tidal wave, 66, 67. *See also* Underground water; Water Supply and Irrigation Papers

Guyer, Michael F.: research on inheritance by, 164; mentioned, 143

Habicht, Professor: scholarships for American students at Swiss universities and Swiss students at American universities, 156

Hagen, Oskar (Department of Art History), 126

Haight, George I.: letter from CSS, 36; active in establishment of WARF, 176; charter of WARF, 182; president of WARF, 190; mentioned, 37

Hall, D. C.: letter from CSS re football coach and Athletic Association, 110–11

Hallmans (Swiss-Canadian relatives of CSS), 14

Hamlin, Homer: technical advice from CSS to, 90

Hanks, Lucien M. (Louis): letter to CSS concerning effect of intercollegiate athletics, 103; active in establishment of WARF, 176; Articles of Organization of WARF, 182; secretary-treasurer of WARF, 186

Hart, Edwin B.: appointed to University Research Committee, 140; grant for histological studies of rickets to, 179

Hart, Walter, 45, 195

Harza, B. F., mentioned 71, 78

Haskins, Charles H.: tribute to Dean Birge by, 146; member Town and Gown, 214

Health problems of CSS, 233

Hefty, Tom, 156

Henry, William A., 134

Hess, Ralph H.: appointed to University Research Committee, 140

Hiestand, William Dixon (Registrar), 150

High, James L.: name suggested for dormitory, 117

Hisaw, Frederick L.: research in hormones by, 159

Hoard, William D., 173

Hobbs, William H.: paper on meteorite, 68; helped by CSS, 69

Hodgkins, Grace Merrill: letter re Northland College to CSS from, 239; mentioned, 267

Hodgkins, Walter (Regent), 267

Hogle, C. E.: in charge of Garden City project, 83

Hohlfeld, Alexander R.: re Carl Schurz Memorial Professorship, 169, 170; mentioned, 102

Holcomb (a farmer near Garden City), 74, 79

Holgate, Thomas F., 17

Holt, Frank O.: refectory named for, 118

Hood, Mamie (high school classmate of CSS), 15

Hoven, Matthew J., 205

Huber, Catherine (married Jacob Bechtel Slichter), 11

Hubers (Swiss-Canadian relatives of CSS), 14

Hutchins, C. P.: letter from CSS concerning baseball schedule, 111–12; mentioned, 112

Industrialism: CSS relates it to science, 202, 203

Ingersoll, Leonard R.: on Fry's examining committee, 32

Ingraham, Katherine: drawing for title page of Science in a Tavern by, 227

Ingraham, Mark H.: appointed to Mathematics department, 1919, 45; feeling for CSS, 53; correspondence with CSS re deanship of, 163; chairman of committee on computer science, 186, 187; letter from CSS asking comment on paper for Madison Literary Club, 225; letter from CSS concerning title for book of essays, 227; chairman, CSS memorial committee, 247

Intercollegiate athletics: effect on University and students, 103

Investments: CSS's stocks and real estate, 199

Jastrow, Joseph: member Town and Gown, 214

Jerome, Harry, 187

Jones, Burr W.: founder of Town and Gown, 214; minutes of Town and Gown meeting, 215–16

Jones, Lewis R.: chairman of committee on improvement of graduate work, 142; letter from CSS on Carl Schurz professorship, 169

Jones, Richard Lloyd: letter from CSS concerning intercollegiate athletics, 107

Jordan, Edward S.: letter from CSS on athletics at Wisconsin, 107; articles written for Colliers' Weekly by, 108

Kansas: underground water, 63; discovery of oil in, 74; dispute with Colorado over use of water, 80

"Karl Schlichtehr's Prophecies," 15

Kiekhofer, William H.: wall of, 130

Kies, William S.: protest against The Daily Cardinal, 113; an organizer of WARF, 141, 176

King, Franklin H.: collaboration with CSS, 56, 57, 68, 69; measuring porosity of sand by, 58; mentioned, 60

Kitchener, Ontario. See Berlin, Ontario

Kohler, Walter: campaign for funds to build Memorial Union, 123

Kowalke, Otto L.: recollection of CSS as teacher, 30, 31, 33; member of Committee on Undergraduate Social Needs, 116

Kraus, Ezra: plant scientist, 143

La Crosse, Wisconsin, water supply: study of, 92, 93; purification by chlorination, 98

La Follette, Philip F.: description of Arizona politics, 172

La Follette, Robert M.: name suggested for dormitory, 117, 118

Laird, Arthur G.: acting graduate dean, 277

Lancaster County, Pa., 9

Lane, Ernest P.: statement on freshman engineering mathematics, 45; mentioned, 45

Langer, Rudolph E.: on CSS as applied mathematician, 52

Laplace equation, 58

Leopold, Aldo: as essayist, 224

Library Hall Committee: CSS member of, 127

Link, Karl P.: comment on CSS as teacher, 34, 158; CSS solicits Brittingham funds for research of, 159–61; interviewed by David F. Allmendinger, 161; comment on CSS as dean, 161

Loevenhart, Arthur: work on syphilis, 175

Lomia, Luigi, 30

Long Island, New York: underground waters, 63; survey of water resources, 69

Lowell, A. Lawrence: comment on standards for master's degrees, 171

Lunn, Arthur C.: paper published by Carnegie Institution of Washington, D.C., 64

McConnell, I. W.: supervising engineer of Garden City Project (1909), 86; on condition of ditches, 86; mentioned, 91

McDowell, Clyde S.: war research at Naval Experiment Station by, 141

McGilvary, Evander B., 102

McGilvary, Paton, 257, 258

Mack, John G. D.: member of committee on dormitories, 114

Mackin, Rosetta (Powers): comment on CSS as teacher, 35; mentioned, 153

MacMillan, W. D.: paper published by Carnegie Institution of Washington, 64; CSS comments on paper of, 66

McNair, Rush: classmate of CSS at Northwestern University, 17

Madison Club: criticized by CSS, 124

Madison General Hospital: proposal to establish, 207

Madison Literary Club: description of, 213–14; CSS's membership in, 214; titles of papers presented by CSS at, 225

Manitoba, Public Utilities Commission: CSS reports to, 94

Manitowoc, Wisconsin: CSS's advice on water supply, 92

March, Herman W.: on Fry's examining committee, 32; praises CSS's *Elementary Mathematical Analysis*, 38–39; career synopsis of, 43; letter from CSS regarding salary increase of, 51; conciliator, 53; member CSS memorial committee, 247; letter to Louis Slichter, re Munich recollections, 262; mentioned, 34, 45, 52, 188

Mason, Max: on Fry's examining committee, 32; backed by CSS for presidency of University of Chicago, 36, 163; asked to recommend candidates for professorship in Mathematics department, 40–41, 44; considered for professorship by CSS, 42–43; degree in mathematics, 55; professorship in physics, 56; Warren Weaver took Ph.D. under, 56, 158; submarine detection and location, 79, 113, 140–41, 176, 240; President, Rockefeller Foundation, 163; appointed president, University of Chicago, 163; suggests CSS for presidency of University of Arizona, 172; mentioned, 37, 40, 42, 44, 45, 52, 141, 143, 163

Mason, Slichter and Hay, firm of, 79

Master's degrees: CSS opinion on granting of, 170, 172; "Debunking the Master's Degree," 170–72

Mathematical Association of America, 31, 39

Mathematics, Department of: CSS as department member, 21–54; number of staff in, 21; CSS as chairman, 39–54; Ph.D.'s in mathematics, 41; appointments under CSS's chairmanship, 45; space problems, 45–50; tensions in, 52

Mathews, J. Howard: initiated National Colloid Symposium, 164

Mead, Daniel: consulting engineer, 92

Mead, Warren J. (geologist): relations with CSS, 157, 158; mentioned, 149

Meiklejohn, Alexander, 119, 122

Memorial Union Building: CSS member "University Committee on the Union," 122; observations from CSS to H. C. Bradley, chairman, concerning, 123; CSS on need for small dining rooms, 124–25 *passim*; memorandum by Porter Butts on, 125–26

Mendenhall, Charles E.: letter to, re overdue reports on graduate students, 166; mentioned, 187

Mendota, Lake: Slichter property on northeast shore, 270; pollution by State Hospital, 271. *See also* Sladshack

Mennonites: migration to Canada, reasons for, 10; mentioned, 8, 9

Merrill, Agnes. *See* Scott, Agnes Merrill

Merrill, Ellen (Byrne), 266

Merritt, Wesley: letters from CSS, concerning geology findings and price of fuel, 72–75; letter from CSS concerning Cimarron River project, 91; *See also* Atchison, Topeka & Santa Fe Railroad

Military instruction: letter from CSS to Governor A. G. Schmedeman, 153–54

Miller, Edward E. (physicist): reference in paper on capillary flow to CSS's work, 60

Miller, William Snow, 166

Millikan, Robert A.: conference on problem of submarine detection, 140

Milton College: CSS's appraisal of, 168

Miner, Carl S., 178

Montgomery County, Pa., 8

Moore, Eliakim H.: compared with CSS, 24

Moots, Elmer E., 45

Moravian Church, 9

Morgan, J. Pierpont: leadership in financial crisis, 83

"Motions of Underground **Waters,** The," 60. *See also* Water Supply and Irrigation Paper, No. 67

Moulton, Forest R.: paper published by Carnegie Institution of Washington, D.C., 64

Munroe, Dana Carleton: signs humorous certificate, 150

Murdock, Victor: letter of congratulations on Garden City project to CSS, 80; mentioned, 81

National Colloid Symposium: originated at Wisconsin, 164

National Research Council: University committee appointed to cooperate with, 137

Nebraska: underground water in, 63

New Dundee, Ontario, 11

Newell, F. H. (United States Geological Survey: Reclamation Service): proposal of, to assist F. H. King's work financially, 57; transmittal letters, 60, 63; correspondence with CSS on problems of Garden City ditches, 84; mentioned, 83

New Hanover, Pa., 8

Newton, Isaac, 20, 66, 226, 228

New York Mathematical Society: CSS a member (1891), 39; predecessor of American Mathematical Society, 39

Nieman, Carl, 161

Nimitz, Chester, 141

Noe, Walter, 211

Northland College: CSS's appraisal of, 167, 168; gifts of books to, 169, 239

Northwestern University: CSS as student at, 16–19

Noyes, George H.: name suggested for dormitory, 118

"Number 9." *See* Research Fund, University

Ochsner, Albert J.: name suggested for dormitory, 118; mentioned, 136

O'Dell, Maria (Byrne), 266

Ogg, Frederick A.: on needs of research in social sciences, 185

Oklahoma-Kansas-Nebraska region: survey of water resources, 69

Olson, Julius E., 29, 30

Osgood, William F.: advice on candidates for professorship in Mathematics department, 42; mentioned, 41

Osterberg, Harold: assistant to Professor Roebuck, 187

Otto, Max: comment on CSS's paper on "Galileo," 230

Owen, Edward T.: minutes of Town and Gown meeting by, 62–63; member of first Committee on Athletics, 104; subscribes to shares for University Club, 130; member Town and Gown, 214; mentioned, 37

Owen, Ray: in field of topographical engineering, 70–71; mentioned, 72, 75, 83

Paper mills: waste from future mills a potential contamination, 96

Parker, C. I. (CSS's high school principal): letter of appreciation from CSS, 14–15

Parkinson, John B., 130

Peabody, Arthur: architectural plans for Memorial Union drawn by, 125

Pence, Nellie Ada (wife of Sumner H. Slichter), 259

Pennsylvania Dutch, 10

Poetter, Reuben S.: letter from CSS concerning hiring extra men, 88; mentioned, 43

Perkasie Meeting House Cemetery, Hilltown, Pa., 8

Peterson, William H.: receives grant-in-aid for studies of bacteria, 185

Phi Beta Kappa: CSS elected to, as alumnus, 16; addresses before, 226

Picnic Point: sand from, used in experiments, 63

Pike, Ann (Slichter): letter to Mark H. Ingraham describing CSS, 294; mentioned, 290

Pike, John: director, State Investment Board, 205

Pitman, Ann, 258

Powers, Rosetta. See Mackin, Rosetta

Preston, Ontario, 11

Professional societies: CSS activities in, 39

Pronleroy, P. de: correspondence with CSS re irrigation proposals, 77

Quakers, 8

Questionnaires: CSS on cost of answering, 167

Railroad Commission of Wisconsin: petition from La Crosse to, 93

Randall, Dean Otis E.: letter from CSS re qualifications for teaching, 27

Reber, Louis F. (Dean of Extension Division), 149

Rechard, Ottis H., 45

Regents, Board of: appropriates $5,000 toward war research work, 140–41; approves appointment of CSS as Dean of Graduate School, 144; resolution to refuse gifts from foundations, 175; rescinds resolution, 175, 176; special meeting of Executive Committee to consider taking out patents, 181

Reinsch, Paul S.: comment on space assignments, 48

Relatives of CSS, 264–67

Religious affiliations of Schlichters, 10

Research Corporation of New York: organized to handle patents for universities and faculties, 178

Research Fund, University: special research fund ("Number 9"), 142; grants from, 155; inadequate for needs, 173–74; balances support of humanities and social studies, 185; significance of, 192

Residence Halls Committee, 116

Retirement of CSS: "The Years of the Granchildren," 289–95

Richards, Harry S. (Dean of Law School), 147

Richardson, Henry L.: name suggested for dormitory, 118

Richardson, R. G. D.: secretary, American Mathematical Society, 24

Richter: meaning of name, 7

Rio Grande valley: survey of water resources, 69

River Falls State Normal School, 56

Robberies of CSS, 235, 236

Robson, H. A.: asks CSS to investigate source of water for Winnipeg, 94; career of, 94; excerpts from editorial on Winnipeg project, 98, 99; letter from CSS re completion of Winnipeg water supply system, 99. See also Winnipeg

Rockefeller Foundation: Warren Weaver's connection with, 56, 68; Max Mason's connection with, 163

Rockford, Ill.: water supply, 92
Roebuck, John R.: research on Joule-Thompson Effect, 187
Ross, Edward A., 102
Royal Philosophers: menu of, 268
Russell, Harry L.: student in Birge's classes, 145; became Dean of College of Agriculture, 1907, 148; activities in WARF, 176, 182, 183, 184, 192; letter from CSS on postdoctorate support, 189; mentioned, 102, 136

Saint Paul, Minn.: CSS born in, 11
Salisbury, O. M.: letter from CSS re University Club finances, 131
Sanborn, Arthur L., 113
Sanborn, John B., 113
Santa Fe Railroad. See Atchison, Topeka and Santa Fe Railroad
Sargent, F. L. (botanist), 29, 30
Schenectady, N.Y.: city water supply, 92
Scheremaier: generalized name used by CSS, 34
Schlichter: meaning of name, 7; common given names, 8; variations in spelling, 8
Schlichter, Christian (great-grandfather of CSS), 8, 11
Schlichter, Daniel (father of Christian): native of Switzerland, 11
Schlichter, John (son of Christian), 9, 10, 11. See also Slichter, John
Schlichter, John, Jr.: pastor of Congregational Church, Sterling, Kansas, 10
Schmedeman, Albert G.: letter about military instruction from CSS to, 153–54; letter from CSS on extension of University Avenue to West Washington Avenue, 209–10
Schuette, Henry, 177
Science: CSS's faith in, 180; related it to industrialism, 202, 203
Science in a Tavern: essays by CSS, 226–27
Scientist: CSS as, 56–69
Scott, Agnes (Merrill): memories of CSS, 254–55; account of European trip, 261–62; mentioned, 260, 263, 267

Sellery, George C.: refuses salary increase for staff member of Mathematics department, 51; requests Birge's advice, 89; reply to Allen Survey, 128–29; relations with CSS, 149, 150; signs humorous certificate, 150; requests University Research Committee to assume salaries, 174; mentioned, 187
Sewage disposal, Madison: CSS's appraisal of, 206–7
Sharp, Frank C., 258
Sharp, Malcolm (son of Frank C. Sharp), 257, 258
Sheehan, Ellen. See Byrne, Ellen Sheehan
Shipman, Mother, 15
Shoal Lake. See Winnipeg: city water supply
Siebecker, Robert G.: name suggested for dormitory, 118
Sigma Chi: CSS member of, 17
Simpson, Thomas M., 45
Skinner, Ernest B.: dinner honoring, 33; sketch of, 40; high school inspections, 127; college texts, 195; mentioned, 23, 39, 44
Sladshack: 270–72; telephone service at, 200–201, 270; fire at, 271; mentioned, 234, 235, 267
Slaughter, Gertrude: reminiscences concerning Slichter boys, 255
Slaughter, Moses S.: high school inspections, 127; member of Town and Gown, 214; memorial to, by CSS, 218
Slichter, Ada (Pence) (wife of Sumner H. Slichter): recorded on "family tree," 290
Slichter, Allen M. (son of CSS): early business enterprise, 201; letter to parents about record album cabinet, 234; birth of, 253; Pelton Steel Casting Company, Milwaukee, 254; arrangement from Homer, 258; boyhood letter to CSS, 258; agrees to keep auto shined up, 259; catalog of art treasures seen in Europe, 260; diary of European trip, 260; letter to parents, re asparagus, 269; military service, 285; wife and children, 290; mentioned, 245

Slichter, Ann (granddaughter of CSS). *See* Pike, Ann (Slichter)

Slichter, Charles P. (grandson of CSS): letter to "Grandma Cakes" (Mrs. CSS), 291; letter to Mrs. CSS after CSS's death, 293–94; mentioned, 258, 290

Slichter, David (brother of CSS), 202, 265

Slichter, Donald A. (grandson of CSS), 290

Slichter, Donald C. (son of CSS): birth of, 253; Northwestern Mutual Life Insurance Company, 254; letter from Agnes Merrill, 257; agrees to keep auto shined up, 259; visit to Louis in California, 269; wife and children, 290; mentioned, 201, 205, 245

Slichter, Dorothy (Doyon) (wife of Donald C. Slichter): recorded on "family tree," 290

Slichter, Dorothy (Fritsch) (wife of Allen M. Slichter): brings record cabinet made by CSS to Allen, 234; recorded on "family tree," 290

Slichter, Elizabeth (Bechtel) (grandmother of CSS), 13. *See also* Bechtel, Elizabeth

Slichter, Frederic (brother of CSS): letters to and from CSS, re evolution, 19, 20; mentioned, 264

Slichter, Jacob Bechtel (father of CSS), 10, 11, 12

Slichter, Jane (granddaughter of CSS), 290

Slichter, John (grandfather of CSS), 11

Slichter, Louis B. (son of CSS): family letters from CSS to, 11, 91, 234, 266, 289, 295; submarine detection and location, 79, 140, 141; letter from Warren Weaver after death of CSS, 158; influenced by Max Mason, 163; birth of, 253; head, Geophysics Institute, UCLA, 254; account of family discipline by, 255; boyhood letter to CSS, 258; agrees to keep auto shined up, 259; diary of European trip, 260; reply to letter from CSS after death of Isabel Byrne, 266; letter to parents, 269–70; letter to Mark H. Ingraham

concerning family transportation, 272–73; letter to CSS re portrait of mother, 282; war work, 285; letter to, from Mrs. CSS about Sumner's wedding, 285, 286; K-W Road Smoothers, 289; wife and children, 290; mentioned, 172, 245, 254

Slichter, Marjorie Ann (granddaughter of CSS), 234, 290

Slichter, Martha (Buell) (wife of Louis B. Slichter): recorded on "family tree," 290

Slichter, Mary Louise (granddaughter of CSS), 290

Slichter, Mary Louise (Byrne) (wife of CSS): letter from Warren Weaver after death of CSS, 158; feeling concerning University of Arizona presidency, 172; buys property on Carroll Street, 196; ancestry, 248; letters to CSS, 251–53, 255; portraits of, 282; letter to Louis about Sumner's wedding, 285, 286; letter to Sumner H. Slichter, 286–88; letters to from Charles P. Slichter, grandson, 291, 293, 294; mentioned, 203. *See also* Byrne, Mary Louise

Slichter, Sumner H. (son of CSS): habit of using snitched stationery, 79, 281; family letters from CSS to, 79, 101, 189, 233, 234, 265, 271, 279, 291; boyhood letters to CSS, 79, 256; address at Engineers' Day dinner, 204; birth of, 253; leading economist, 254; letter to Mrs. CSS from Switzerland, 268–69; family letter, 269; military service, 285; family letter from Mrs. CSS, 286–88; wife and children, 290; mentioned, 245

Slichter, Susan Merry (granddaughter of CSS): has measles, 11; letter from CSS to, 293; mentioned, 290

Slichter, William P. (grandson of CSS): letter to Sumner H. Slichter from the Philippines, 292; mentioned, 282, 290

"Slichter block": problems in construction and maintenance, 196, 197

Smith, Charles F.: member Town and Gown, 214

Smith, Howard L.: letters from and to CSS concerning funds for prospective athletes, 105; introduced by CSS at dinner for E. A. Birge, 133; mentioned, 102, 132

Smith, Walter M. (University librarian), 132

Soils Management, Division of: Franklin H. King, chief of, 56

Sommerfeld, Arnold: first post–World War I Carl Schurz professor, 170

South, University of the: CSS gives books to, 169, 239

South Platte river and valley, 90

Southern California: flow of underground water, 62

Spooner, John C. (senator): CSS writes to, re alcohol tax, 81; name suggested for dormitory, 117

Sproule, A. H.: letter from CSS re rules governing athletes, 111

State Historical Society of Wisconsin: building to be used jointly with University of Wisconsin, 127

Steenbock, Harry: research in vitamins, 159; the establishment of WARF, 176, 182, 192; biographical sketch, 177; irradiation of foods, 177, 178; faculty memorial to, 177–78; first application for patent, 179; subsidizes publication Science in a Tavern, 226; funds for financing first foreign patent, 240; member CSS memorial committee, 247

Steiglitz, Julius: paper published by Carnegie Institution of Washington, D.C., 64

Sterling, John W.: taught first class at University of Wisconsin, 22

Stevens, Breese J.: member Town and Gown, 214

Stevens, L. E. (Madison realtor): remarkable stationery of, 198

Student Union. See Memorial Union Building

"Suggestions for the Construction of Small Pumping Plants for Irrigation," 89. See also Water Supply and Irrigation Paper, No. 184

Svedberg, The (professor from Upsala): participates in first National Colloid Symposium, 164

Swenson, Magnus: letter to CSS concerning effect of intercollegiate athletics, 102; subscribes to shares for University Club, 130

Swiss Brethren. See Mennonites

Swiss universities: scholarships for American students at, 156

Tally, R. E.: contacted CSS re presidency of University of Arizona, 172

Tarrant, Warren D.: name suggested for dormitory, 118

Taylor, Eugene, 45

Taylor, Robert: account of early days of WARF by, 179

Teaching of CSS, 22–37

Teaching profession: CSS's opinion of importance of, 171

Texas: underground water, 63

Textbooks: CSS's writing of, 38, 92, 194; CSS collaborates with Van Velzer, 195

"Theoretical Investigation of the Motion of Ground Waters." See Ground water

Thompson, Silvanus: author of Calculus Made Easy, 39

Thorkelson, Halston J., 148

Thwaites, Frederik T.: assistant to CSS on Garden City project, 72

Tottingham, W. E.: K. P. Link received Ph.D. under, 158; mentioned, 187

Town and Gown: 214–21; minutes quoted, 63, 215, 216, 217, 218, 264, 277–78; historical background of, 214–15, 220–21; CSS's memorials to Slaughter and to Turner, 218–19; mentioned, 62, 235

Townley, Sidney E.: comment on CSS as teacher, 29

Travel: CSS's fondness for, 276–80

Tripp, J. Stephens: estate of, 115, 116; name suggested for dormitory, 117

Tripp Hall: men's dormitory, 116, 117

Trumbower, Henry R.: recalls class incident, 34; member CSS memorial committee, 247

Tupper, W. S., 30

Turneaure, Frederick E. (Dean of College of Engineering): member of committee on dormitories, 114; career of, 147

Turner, Frederick J.: reform of intercollegiate athletics, 104; committee to write constitution for intercollegiate debating team, 127; member Town and Gown, 214; memorial to, by CSS, 219; letter commenting on CSS's Founder's Day speech in Chicago, 241–43; mentioned, 30

"Twenty, The": Mennonite settlement, 9

Underground water: CSS's chief scientific work connected with, 56; movement of, 62; economically obtained, 70; problems in extensive withdrawal, 85; CSS services as consultant on, 194, 195. See also Ground water; Water Supply and Irrigation Papers

Union. See Memorial Union Building

United Brethren, 9

United States Civil Service Commission: letter of protest from CSS, 71

United States Geological Survey: CSS's scientific papers published in annual reports, 57; CSS hydrographer for, 69; mentioned, 28, 56. See also Water Supply and Irrigation Papers

United States Reclamation Service: CSS's services to, 69, 88; offers to increase capacity of plant at Garden City, 86

University Avenue: CSS's opposition to extension to West Washington Avenue, 209, 210

University Club: criticized, 124; CSS's connections with, 129–32; rules for steward at, 131, 132

University of Wisconsin, Faculty of: memorial resolution on CSS, 245–47

University of Wisconsin Alumni Association: CSS's speech to, 134–37

University of Wisconsin Athletic Association: membership of, 106

University of Wisconsin Foundation: free fund supported by annual gifts, 183

University Research Committee: 139–43; to secure information on research projects, 142; methods of handling funds, 165; action on request to assume some salaries, 174

Van Hise, Charles R.: advice from Ford on candidates for professorship in mathematics, 41; decision to offer professorship to Van Vleck, 43; requests information on space and space needs, 45; member of first Athletic Council, 105; secures funds for women's dormitory, 114; administration building named for, 118; name suggested for dormitory, 118; refectory named for, 118; first promoter of Union Building, 122; subscribes to shares for University Club, 130; opinion that University must become best of all, 135; announces cooperation with National Research Council, 139; appoints special committee on improvement of graduate work, 142; receives first Ph.D. degree at University of Wisconsin, 157; member Town and Gown, 214; letter from CSS in Munich, 263; mentioned, 44, 49, 50, 102, 108, 109, 110, 113, 144, 145, 147, 267

Van Hise, Janet (daughter of C. R. Van Hise), 118

Van Velzer, Charles A.: resigns as chairman of Mathematics department, 39, 40; teaching Letters and Science students, 45; coauthor with CSS on textbooks, 68, 195; mentioned, 21, 22, 23, 43

Van Vleck, Edward B.: dinner honoring, 33; considered for professorship in Mathematics department, 42; letter to CSS accepting professorship, 44; contrasted with CSS, 52, 53; comment on *Science in a Tavern*, 53; mentioned, 24, 27, 45, 48, 188

Van Vleck, John Hasbrouck (son of E. B. Van Vleck), 43

Van Vleck, John Monroe (father of E. B. Van Vleck), 42

Vilas, William F.: name suggested for dormitory, 117; mentioned, 124, 133

Wanderbach, Mary (wife of Christian Schlichter), 11

War research work at University of Wisconsin, 140

WARF. *See* Wisconsin Alumni Research Foundation

Warren, Glenn B.: General Electric Company, 35–36

Washington and Lee University: CSS gives books to, 169, 239

Waterloo County, Ontario, 9, 10, 11

Water Resource Board: history of, by Robert Follansbee, 62

Water supplies, city: 92–100; advice of CSS sought on, 92. *See also* Brooklyn; La Crosse; Manitowoc; Rockford; Schenectady; Winnipeg

Water Supply and Irrigation Papers (United States Geological Survey):

—No. 67 – "The Motions of Underground Waters," 60;

—No. 140 – "Field Measurements of the Rate of Movement of Underground Waters," 63;

—No. 187 – "Suggestions for the Construction of Small Pumping Plants for Irrigation," 89

Water table: relation to flow in wells, 59

Waters, Ralph M. (anesthetist), 148

Weaver, Warren: comments on CSS, 33, 34, 165; distinguished student of CSS, 52; warm regard for Van Vleck, 53; degree in physics, 56, 158; chairman of Mathematics department, 56, 158; helped secure Svedberg ultracentrifuge, 68; close relationship with CSS, 158; vice president of Rockefeller Foundation, 158; letters to Mrs. CSS and to Louis Slichter after death of CSS, 158; influenced by Max Mason, 163; member Town and Gown, 217; mentioned, 37, 45, 157, 188

Western Electric, 31

White, Helen C.: appreciation of CSS, 165

White, Henry S.: considered for professorship, 42, 43

Whitewater State Normal School, 57

Whittaker, Edmund T., 41, 42

Wild flowers: CSS's appreciation of, 270, 272

Wildlife: CSS's appreciation of, 275

Williams, Bill: stories of his dog, 237–39

Wilson, Perry W.: research in nitrogen fixation, 161; letter of commendation from CSS, 161, 162; reply to CSS, 162; mentioned, 158

Winnipeg: city water supply, 92, 93–100; earlier study by special commission, 93–94; Robson employs CSS as expert, 94; CSS reports on his study, 94–98; project completed, 98; editorial on, 98–99; CSS writes Robson re satisfaction with, 99–100

—Shoal Lake, as source of city water supply for, 93, 95, 96, 97, 98, 100; re temperatures of lake waters, 95

—Winnipeg River, as source of city water supply for, 93–94, 95, 96, 97, 98

Winslow, John B.: dormitory house named for, 118; nominated by CSS for membership in University Club, 131; member Town and Gown, 214

Winterbotham, J. W.: letter from CSS re gift of stamps to sons, 259

Wisconsin Academy of Sciences, Arts and Letters: CSS president of, 39

Wisconsin Alumni Research Foundation: 176–93; funds for scholarships, 154; established, 176–83; Trustees of, not on University staff, 183; CSS role in spending funds from, 184; significance of funds, 192; tribute to CSS, 244; mentioned, 110, 113

Wisconsin Salon of Art: started by Porter Butts, 126

Wisconsin State University, River Falls. *See* River Falls State Normal School

Wisconsin State University, Whitewater. *See* Whitewater State Normal School

Wisconsin Telephone Company: CSS's troubles with, 200–201

Wolff, Henry C.: Ph.D. thesis under CSS's direction, 23; applied mathematics, 40; Garden City project, 70, 71; career synopsis, 71–72; memorandum from CSS regarding tests to be made in Colorado and Nebraska, 88; letter from CSS concerning firing men, 89; mentioned, 39

Yost, Fielding, 107